OUT OF THE ZENITH

OUT OF
THE ZENITH
Jodrell Bank
1957–1970

BERNARD LOVELL

HARPER & ROW, PUBLISHERS

NEW YORK, EVANSTON, SAN FRANCISCO, LONDON

FIRST U.S. EDITION

ISBN: 0-06-012719-8

LIBRARY OF CONGRESS CATALOG CARD NUMBER: 73-14269

CONTENTS

LIST OF PLATES

following page 116.

1. The author and Professor J. G. Davies explaining the results of the observations of the Soviet probe to Venus

2a. The Mk II radio telescope at Jodrell Bank

2b. The Mk III radio telescope at Wardle near Nantwich, Cheshire, controlled by radio link from Jodrell Bank

3. The Lick Observatory stroboscopic television photographs taken on 3 February 1969 of the light pulses from the Crab nebula pulsar PSR 0531+21

4a. Dr J. E. B. Ponsonby and the 'ephemeris doppler machine'

4b. Sir Martin Ryle and Professor Graham Smith

5. The planetary system as drawn by J. G. Davies (*top*) and as received (*bottom*) at Zimenki Observatory in the U.S.S.R.

6. The receiving laboratory at Zimenki Observatory in the U.S.S.R. during the Jodrell Bank–Echo II–U.S.S.R. communication experiments in February/March 1964

7. The Mk I radio telescope and the Mk IA

8. The Mk I telescope during the conversion to the Mk IA, photographed on 20 April 1971

The plates have been reproduced by permission of the following: Pl. 1 *The Guardian*; Pl. 3 Lick Observatory; Pl. 4a C.O.I.,; Pl. 4b Mullard Ltd.; Pl. 6 Tass; Pl. 8 Airviews Ltd., Manchester Airport. Plates 2a and b, 5 and 7 are from Jodrell Bank.

PREFACE

Although this book is concerned with astronomy and radio astronomy I must make it clear that I have made no attempt to write a textbook. It is essentially a book about Jodrell Bank—and indeed a localized part of Jodrell Bank, concerned with the researches carried out with the 250 ft aperture steerable radio telescope (the Mark I) from the moment when work began with it in 1957 to the time thirteen years later when operations ceased for several months for renovation and changes converting it to the Mark IA. During these years astronomical discoveries of immense significance have been made in many parts of the world. The fact that some are covered here and others neglected reflects only the extent to which the Mark I telescope has been involved. I offer this explanation and apology to my friends and colleagues elsewhere in the world who may feel, as one of them remarked on a previous occasion, that for me the universe is centred on Jodrell Bank.

I am especially indebted to Dr. H. P. Palmer for his detailed advice on Chapters 2–6 which describe the researches in which he has been so deeply involved; and similarly to Dr. A. G. Lyne in connection with Chapters 10–12 on pulsars. Also I particularly wish to thank Professor W. H. McCrea, F.R.S. for his advice on Chapter 9, and Professor F. G. Smith, F.R.S., and Professor J. G. Davies for their advice and help with this account of the researches with which they have been so closely associated.

BERNARD LOVELL
Jodrell Bank
1970–73

ACKNOWLEDGEMENTS

The author and publishers wish to thank the following who have kindly given permission for the figures quoted to be based on diagrams first appearing elsewhere.

Figs. 1, 2, 3, and 4: Taylor and Francis Ltd., R. Hanbury Brown *et al.*, *Phil. Mag.*, 7, **46**, 857, 1955; Fig. 5: Stanford University Press, 'The Relation of Radio Astronomy to Cosmology' by F. Hoyle in *Paris Symposium on Radio Astronomy* ed. by Ronald N. Bracewell (1959), p. 532; Fig. 6: O. Elgaroy *et al.*, *Mon. Not. R. astr. soc.*, **124**, 395, 1962; Figs. 7b and 8a and b: B. Anderson *et al.*, *Nature*, **195**, 165, 1962; Fig. 9: N. W. Broten of the Algonquin Radio Observatory (National Research Council of Canada) *et al.*, *Mon. Not. R. astr. soc.*, **146**, 313, 1969; Fig. 10: G. K. Miley, *Mon. Not. R. astr. soc.*, **152**, 486, 1971; Fig. 11: R. D. Davies, Jodrell Bank; Figs. 12, 13, 14 and 15: R. J. Peckham, Jodrell Bank; Fig. 20: Columbia University Press, *Galaxies and the Universe* by I. Woltjer (1968), p. 104; Figs. 21 and 23a: University of Chicago Press, *Astrophys. J.* **133**, 355, 1961; Fig. 22: *Q. J. R. astr. soc.*, **13**, 282, 1972; Figs. 24, 25, 26a and b, and 27: J. G. Davies *et al.*, *Nature*, **217**, 910, 1968; Fig. 28: B. J. Rickett, *Nature*, **221**, 158, 1969 and *Mon. Not. R. astr. soc.*, **150**, 67, 1970; Fig. 29: B. J. Rickett, *Mon. Not. R. astr. soc.*, **150**, 67, 1970; Fig. 30: A. Hewish *et al.*, *Nature*, **217**, 709, 1968; Fig. 31: J. G. Davies *et al.*, *Nature*, **221**, 27, 1969; Fig. 32: V. Radhakrishnan *et al.*, *Nature*, **221**, 443, 1969; Fig. 33: V. Radhakrishnan and R. N. Manchester, *Nature*, **222**, 228, 1969; Fig. 34: P. E. Reichley and G. S. Downs, *Nature*, **222**, 229, 1969; Fig. 35: A. G. Lyne and B. J. Rickett, *Nature*, **218**, 326, 1968; Fig. 36: J. G. Davies *et al.*, *Nature*, **240**, 229, 1972; Figs. 37 and 38: R. R. Clark and F. G. Smith, *Nature*, **221**, 724, 1969; Fig. 39: G. de Jager *et al.*, *Nature*, **220**, 128, 1968; Figs. 40 and 41: W. J. Cocke *et al.*, *Nature*, **221**, 525, 1969; Fig. 42: T. Gold, *Nature*, **221**, 25, 1969: Fig. 43: Lund Observatory; Figs. 44 and 45: C. Hayashi, *Publ. astr. soc. Japan*, **13**, 450, 1961; Fig. 49: *Q. J. R. astr. soc.*, **4**, 347, 1963; Fig. 50: J. E. B. Ponsonby, Jodrell Bank.

OUT OF THE ZENITH

I

The Mark I radio telescope

At 09 00 hours B.S.T. on 14 August 1970 the 250 ft reflector of the Mark I radio telescope at Jodrell Bank was driven to the zenith for the last time. The researches which began thirteen years earlier with this instrument were at an end. It could have been a moment of sadness and farewells, but it was not, because engineers were already on site to begin the process of modification and changes which by 1971 would re-create the telescope as the Mark IA—an instrument looking only marginally different to the casual observer but nevertheless a much improved telescope with stresses and strains removed to give it a further long life of research.

Thirteen years is a long time in the life of a modern scientific instrument. Multi-million-pound devices of our age can become obsolescent in a decade—in space research within a few months—and can be considered to have justified their cost. The Jodrell telescope was the first of its kind and it seems to me, now, altogether remarkable that after thirteen years of almost continuous work it remained the largest, still working on a twenty-four-hour basis, and leaving an unfilled gap in astronomical researches during the months of renovation.

The story of the conception and the building of this telescope and the drama which surrounded the beginning of its work in 1967 is already on record.* The telescope came to life and Jodrell Bank survived by the narrowest of margins. Was it worth while, and how does one judge the success or otherwise of a project like this? There are no profits in the conventional sense—nothing akin to the number of millions of passenger miles flown to be equated against the capital cost of an aircraft, for example. Clearly there is no basis of comparison with any conventional financial enterprise. In the beginning those involved did not believe that the telescope was being built for profit of this kind. But they did believe that it would make a major British contribution to man's knowledge of the universe. They believed that its work would stimulate and encourage the striving of man to understand the universe and his place in it.

* Bernard Lovell, *The Story of Jodrell Bank* (London and New York, 1968).

And how can one place a cost on the worthwhileness of such ambitions for, 'without a vision the people perish'?

In this short pause with the telescope at rest I feel impelled to give some account of my stewardship of that unique instrument. I do this for two reasons. First of all, although I have no doubt about the answers to the questions posed above, I readily concede that others should be the ultimate judges in these enterprises where the profits are of the mind and not the factory. Second, when I reflect on these years I am amazed at the directions in which our researches have turned, and moreover on the strange and frequently chance circumstances which have stimulated these changes. Perhaps it is true that in some scientific research it is possible to programme and plan according to a rigid and agreed scheme. I have not yet encountered many such possibilities in fundamental research. Indeed if I had been confined to the programmes of research on which I had based my case for building the telescope there would be little to write about now. The instrument survived as a major feature of the scientific world because of the complete freedom which I enjoyed with my small group of colleagues, to use it for whatever purpose we wished—always assuming, of course, that I kept within the finances made available to me by a generous university.

In the beginning the University, the Nuffield Foundation, and the Department of Scientific and Industrial Research set up a 'Board of Visitors' with representatives from those three bodies, as well as the Royal Society, and the Royal Astronomical Society. The purpose of this Board was to make sure that I did not misuse the instrument, and to settle any conflicting claims for its use which might arise from external requests. The Board met on two occasions, I think, in the early years and then, apparently by unwritten mutual consent, has never assembled again. There has, in fact, never been conflict of this kind. I have been asked occasionally by outside governmental agencies, both British and foreign, to give assistance with the telescope, but never instructed. I have responded to most of these requests—but not to those asking me to join in the search for extra-terrestrial life—with pleasure. They have involved the telescope in some, but by no means all, of the 4877 hours of 'miscellaneous use'—7·1 per cent of its operational time.* Of that miscellaneous use, 2498 hours has been directly concerned with the space programme of the U.S.A. and the U.S.S.R.—this 3·6 per cent of the use of the telescope is the main reason for which it is known to the public. The hidden values in this work have been tremendous. It has been my good fortune to work in a situation where I had the freedom to engage in these

---

* From August of 1957 when we began preliminary calibration measurements to August 1970 the telescope gathered results for 68 538 hours.

activities; the life of Jodrell Bank would have been more ordinary and mundane without these excursions into fields of activity which were not the primary reasons for our existence.

The conflicts which have occurred have been entirely internal. It seems extraordinary that normally unemotional colleagues have shown passion and anger about their allocation of time on the Mk I. Groups who may already have been using the telescope day and night for two or three weeks have frequently descended on me in force to argue and plead for another twenty-four hours. In thirteen years the occasions on which I have had difficulty in filling even the odd day have been few and far between. In innocence I imagined that with the advent of the smaller but more modern and accurate Mk II and III telescopes in 1964–6, the pressure on the Mk I would decrease. In fact the contrary happened. Investigations of great interest soon began in which either the Mk II or III were used in combination with the Mk I as an interferometer, and the task of telescope allocation became harder than ever—particularly in the later computer phases when on-line working of limited computer capacity was demanded. Always when the pressure might have eased another discovery suddenly emerged, such as quasars or pulsars, which, as if by some miracle, again made the Mk I into a front-line instrument. Even when the Mk I finally went to the zenith in mid-August 1970, an experiment on pulsars was begun to use the instrument in this fixed zenithal position for as long as the contractor's erection programme allowed.

Even if the Board of Visitors to the telescope never had matters of urgency or argument to settle, their existence had the fortunate effect of stimulating me to keep records of the amount of telescope time spent on various researches. During the last three years of its use the telescope was at work on research, that is, with wheels turning, for 90 per cent of the available time. I do not know what the equivalent figure is for comparable large machines in other branches of science, but I suspect that this figure is extraordinarily high. It is a tribute to the small engineering staff whose job it has been to do the daily maintenance on the telescope and to the scientific staff who have been driven to make their equipment so that it is reliable. In this connection I suspect, although I cannot produce statistics to prove this, that a significantly higher rate of serviceability for 1965 onwards is related to the increasing use of transistors, printed circuits, and parametric amplifiers, and the rapid disappearance of the thermionic valve from the research equipment.

The achievement of these high working efficiencies in the general framework of freedom presents peculiar problems. The maintenance of a complex and massive instrument and the safety of the people working with it demand rigid organization and discipline. These attitudes are anathema to the researcher, who insists that he should be free to work when necessary and desirable, be it midday or midnight. I cannot claim to have reached a perfect solution, and as the complexity of the researches has increased so has the difficulty of maintaining the necessary equilibrium. Freedom to choose certainly does not mean that on the basis of some whim one can suddenly switch around the research programme in the middle of the night. The essential requirement is the localized rigidity within the longer term framework of choice.

I make no attempt in the chapters which follow to give a detailed account of the total research output of the telescope. I have chosen those aspects of the work which illustrate the remarkable effects of the freedom of the academic scientist working in this environment where it is possible to react instantly to unexpected influences. I have omitted any detailed description of at least one third of the research output of the telescope. Apart from the bearing on technical developments described in Chapter 8, I have not dealt with the observations of the emission of the 21 cm spectral line from neutral hydrogen. The effect of the discovery of this radio emission on the original design of the Mk I has already been described in *The Story of Jodrell Bank*. Subsequently, the instrument was used extensively in the observation of this spectral line and R. D. Davies with his group has produced a series of results on the distribution of hydrogen in the Milky Way, and in extragalactic nebulae, of fundamental importance to the development of these studies. However, these researches, although not lacking in importance compared with those described in this book, have not been subject to the sudden changes and influences which have characterized the work on radio galaxies, quasars or pulsars, for example. The many sky surveys, and the investigation of the polarization of the radio emission from the radio galaxies and quasars by R. G. Conway and his group have deepened our insight into the mechanism by which the radio waves are generated, but again these researches have followed a calmer course of planned observations than those I have chosen to describe in the following chapters.

2
Radio galaxies

In the memorandum (the 'blue book') which I wrote in 1950–1 to support the final application for the money to build the telescope, there was a chapter setting out proposals for the use of the instrument on the basis of our knowledge about the relation of the radio emissions to the universe. Under the general heading '(2) Galactic and extragalactic radio emissions' there was an item (c) which read 'A complete survey of the position and distribution of the radio stars could be made, yielding crucial information for the establishment of the nature of these sources and any relation with known astronomical objects'.

When I wrote those words I could have had no idea that the use of one technique in the pursuit of this ambition would involve the telescope in 17 600 hours of work, forming a quarter of its entire research programme. In fact the use of the word 'radio stars' is the indicator that at that moment there were few who doubted that the phenomenon of the radio emissions from space was effectively confined to our own galaxy. The ingenious and nearly simultaneous development of two different types of radio interferometer had demonstrated the existence of localized or discrete sources of radio emission amongst the general diffuse background. In Australia the group under the late J. L. Pawsey had mounted a small aerial system on a cliff overlooking the sea and by virtue of using the sea as a reflector had produced the radio equivalent of Lloyds mirror interferometer. The baseline of the interferometer was equal to twice the height of the aerial above the sea, the interference pattern being generated by the interaction of the wave received directly from the source and the wave reflected by the sea. In June 1947, J. G. Bolton, and G. S. Stanley detected the interference pattern from the radio waves emanating from the region of Cygnus as it rose through the aerial beam. They were able to conclude that the radio waves must be emanating from a localized source in Cygnus whose angular diameter was less than 8 minutes of arc.

This was a most interesting and important confirmation of the conclusion reached more than a year earlier by J. S. Hey, S. J. Parsons, and

J. W. Phillips in their pioneer investigations of the radio emissions from space using ex-army radar receivers. They had observed that although the radiation from the sky was, in general, steady, that from the region of Cygnus exhibited peculiar fluctuations with periods of about one minute. They concluded that there must be a variable source of radio waves with small angular diameter in that region. Ironically, it transpired eventually that they had reached the correct conclusion although the radiation from Cygnus is steady. The fluctuations observed by Hey and his colleagues are a result of the transmission through the ionosphere—an effect analogous to the twinkling of ordinary stars. Nevertheless, their conclusion that the Cygnus region contained a radio source of small diameter stimulated the Australian development which led to the unambiguous confirmation.

Meanwhile, in the northern hemisphere, Martin Ryle and Graham Smith were beginning observations with a different type of interferometer—the radio analogue of Michelson's optical stellar interferometer. In this, the radio waves received in two aerials some distance apart on land were fed to a common receiver. The polar diagram of the device is a system of lobes. The angular width $\delta\theta$ of a lobe is determined by the wavelength λ and the separation of the aerials d, and is given simply by $\delta\theta = \lambda/d$. If the radio waves from space emanate from an extensive region then as the rotation of the Earth sweeps the lobe pattern over the sky the signal obtained will remain steady. On the other hand, if the source has an angular diameter smaller than or comparable with $\delta\theta$ then maxima and minima in the signal strength will be observed as the Earth rotates. Furthermore, by varying the spacing (or λ) and hence $\delta\theta$ a measurement of the diameter of the emitting region may be obtained. As an example, if two aerials are separated by 3·5 km and the wavelength is 1 m then $\delta\theta$ is about 1 minute of arc.

By using an interferometer of this type in Cambridge, Ryle and Smith in 1948 confirmed the existence of the source of small diameter in Cygnus and discovered an even more intense source in the constellation of Cassiopeia. The remarkable feature of these discoveries and of those immediately following was that no correlation was apparent between these newly discovered radio sources and known optical objects such as bright stars or nebulae. Only in one case was there a significant correlation, namely the source discovered by the Australians in Taurus, which seemed to be satisfactorily associated with the Crab nebula—the expanding gaseous shell of the supernova of A.D. 1054.

At the time when I was pleading the case for the telescope only a dozen or so of these small-diameter sources had been located and many astronomers believed that the radio emissions were entirely localized in the Milky Way and that these unidentified small-diameter radio sources really were 'radio stars', that is, a hitherto unknown type of dark object in the Milky Way capable of emitting radio waves, but not visible even in the large optical telescopes. Indeed there were very few astronomers at that time who believed that radio waves from outside the galaxy formed any part of the radio phenomena. Those who did so, were soon to be proved right in dramatic fashion.

First of all, R. Hanbury Brown and C. Hazard at Jodrell Bank used the 218 ft fixed transit radio telescope to study the emission from the region of sky containing the M31 extragalactic spiral nebula in Andromeda. Their results, published in 1950, showed unmistakably that M31 was a radio emitter and that although the signals received on Earth were extremely weak, when account was taken of the distance of M31— 2 million light years—it appeared that the radio-emitting properties of M31 were similar to those of the Milky Way. The proof that this extragalactic nebula, and soon afterwards a number of others, was also radio emitting, did not solve the problem of the unidentified small-diameter sources discovered by the use of the interferometers. Those were strong compared with the signals from the normal extragalactic nebula.

The difficulty was that by conventional astronomical standards these early radio measurements did not give really precise positions for these localized sources. For example, they defined only that the Cygnus and Cassiopeia radio sources lay somewhere within a region of sky extending around a degree or so. The existing sky atlases did not show any unusual objects in those regions of the sky—only large numbers of faint stars. In order to justify a detailed search with the world's largest optical telescope—the 200 inch on Palomar Mountain—a far more accurate radio position was needed. Graham Smith solved this problem. He measured the radio positions of the Cygnus and Cassiopeia sources so that the ambiguity in their location was reduced to only a minute of arc. At last it was feasible to make a detailed search of the possible regions of sky with the 200 inch telescope. In 1951 W. Baade and R. Minkowski* made the historic announcement that they had been able to obtain photographs of the objects responsible for the radio emissions in both cases. In Cassiopeia they found faint gaseous filaments which they identified as a hitherto unknown supernova remnant in the Milky Way.

In the position of the Cygnus radio source their long exposure photograph revealed a faint object of a type never before recognized. From the spectrum of the object they estimated that it was exceedingly

---

* The plates were obtained in September 1951. The full account of the work was published by Baade and Minkowski in *Astrophys. J.*, **119**, 206, 1954.

remote—700 million light years; and from a study of the photograph they suggested that this was a case of two galaxies in collision, in which the interstellar gas of the galaxies would be in a high state of excitation— as was, in fact, indicated by the spectrum.

Within a few years other similar identifications followed and thus was born the concept of the radio galaxies—the belief that the small-diameter unidentified radio sources were not radio stars in the galaxy, but on the contrary peculiar objects—believed then to be galaxies in collision—at great distances in the universe, possessing the strange property of emitting radio waves of such strength that they could be detected on Earth, but whose light output was too small to be recorded by the world's largest optical telescope.

It is a remarkable fact that in 1957 when the Jodrell Bank radio tele-scope came into use, only 8 of the discrete small-diameter radio sources had been identified optically, although by that time several thous-and radio sources had been revealed in the radio surveys of the northern and southern hemisphere. Of these 8, three were in the Milky Way (Cassiopeia, Taurus, and Puppis), one was the M31 Andromeda nebula and the other 4 (Cygnus, Perseus, Virgo, and Centaurus) were remote objects in the universe, believed to be colliding galaxies.

Hanbury Brown and the new interferometer

One day early in May 1949 when we were struggling with the early con-cepts of the large steerable telescope I received a telephone call which was to have a vital effect on our future. It was from my friend and col-league, F. C. Williams, Professor of Electrical Engineering in the Uni-versity. He asked me if I remembered Hanbury Brown with whom we had worked at various stages of the war. Of course I did; although he had gone to America in 1942 Hanbury Brown was firm in our memory as the young man in the lab coat who had wielded the pruning knife with such efficiency that our neighbour had assumed he was the gar-dener.

After the war Hanbury Brown returned to England and joined the consultant firm of Sir Robert Watson Watt and Partners Ltd. The mes-sage which Williams conveyed to me over the phone was that he had recieved a letter from Hanbury Brown who wished to explore the pos-sibility of returning temporarily to a University to do research for a Ph.D. He generously suggested that my developments at Jodrell might be of more interest to Hanbury Brown than his own work (at that time he was developing the first of the major computers).

The suggestion seemed like a gift from heaven. At that time Hanbury Brown was one of the most experienced electronic engineers in the country. Before and during the war he had been associated intimately with the development of airborne radar systems first in England and then in America at the Naval Research Laboratory. However, the translation of the phone call to the actual situation of Hanbury Brown working at Jodrell Bank remained a practical problem. The first step was to find out if he was interested in Williams's idea. I met him on the afternoon of 19 May 1949 at the local Goostrey railway station. Before we were off the platform I realized that he had not made an idle inquiry. He really did wish to return to research and the problem was mine only—namely to find him a research post. At that time University posts were hard to get—the expansion of the universities had not begun. Furthermore, Hanbury Brown was already thirty-three with no academic experience and, because all his research had been secret, no published work.

The only immediate possibility was to make an application for one of the I.C.I. Research Fellowships and the Committee was about to meet. R.H.B. to B.L. 24 May: 'I was very impressed by your experimental station and am very keen to join you. I think it is just the right thing. . . . I regret that my decision to come must depend on the Fellowship and if it is offered I shall certainly come. Let us hope the application is successful.' Blackett* had been impressed by my account of Hanbury Brown's potential value to our work and he and Williams, backed by the comments of external referees, overcame the scruples of the Fellowship Committee who were apprehensive about the appointment of a non-academic person of that age to a Fellowship. B.L. to R.H.B. 2 June 1949: 'I was delighted to hear from Blackett yesterday that you had been awarded an I.C.I. Fellowship. . . . as far as we are concerned, the sooner you come the better.'

Although the general idea of Hanbury Brown joining us at Jodrell Bank had been settled, the question of his actual research problem as a Ph.D. candidate remained. At that time we were working on many problems in radio and radar astronomy. B.L. to R.H.B. 20 June 1949: 'When you come up next week we must discuss what sort of work you might like to do . . . for example, there is a great harvest to be reaped in the cosmic noise field when we can bring some intelligence to bear on it.' 6 July: 'I spoke to Blackett on Sunday evening about the proposal that you should come in on the cosmic noise programme and he was most enthusiastic. There is no doubt a great future in this work and I am delighted that you approve of the idea'.

* P. M. S. Blackett, subsequently Lord Blackett, O.M., C.H., F.R.S., was then Langworthy Professor of Physics in the University of Manchester. His vital influence on my career and on Jodrell Bank has been described in *The Story of Jodrell Bank*.

Hanbury Brown settled in during September of 1949. His impact was instantaneous. Just over a year later my annual report to the University Registrar on the I.C.I. Fellow read as follows: 'November 1950 *R. Hanbury Brown*. Has spent the first year in studying the extraterrestrial radio frequency emissions using a narrow pencil beam aerial system at the Jodrell Bank Experimental Station. His work has led to the very important discovery of the radio emissions from the extragalactic nebula in Andromeda.'
Another year:

November 1951. During the first two years of his tenure of the I.C.I. Fellowship Hanbury Brown has carried out an investigation of galactic and extragalactic radio emissions with the large radio telescope at Jodrell Bank.* He has succeeded in obtaining a series of most important results. In the autumn of 1950 radio emissions were measured from the extragalactic nebula in Andromeda, and since then emissions have been measured from four other individual extragalactic nebulae and also from two groups of clusters of nebulae ... A great deal of information about radio emissions from the Milky Way system has also accumulated during the course of this work.

During these two years the discovery of the localized radio sources was emerging as a major problem in astronomy. The circumstances which led to their discovery and the identification of the objects in Cassiopeia and Cygnus have already been mentioned. F. G. Smith had solved the problem of measuring the positions with sufficient accuracy to facilitate the identification of these two sources with the 200 inch Palomar telescope, but the number of unidentified sources soon exceeded a hundred. Neither the Sydney nor Cambridge interferometers had been able to measure their angular diameters. The original Sydney measurements had shown that the source in Cygnus must be less than 8 minutes of arc in angular extent, but further efforts to resolve the sources had only succeeded in slightly reducing these limits.

This question was of fundamental importance. An extent of a few minutes of arc is vast compared with the diameter of the stars—Sirius for example has an angular diameter of only 0·0063 seconds of arc. It seemed clear that before much progress could be made towards an understanding of the radio objects it was essential to find out if they were objects of the same order of angular size as the stars or tens of thousands of times larger, these being the limits of the existing radio measurements.

In principle, to improve the resolving power of the radio interferometer it was merely necessary to increase the separation between the aerials. At that time in the development of the techniques it was not

* The 218 ft diameter fixed transit radio telescope.

possible to do this by any significant extent because the problem of maintaining an adequate stability of phase in the transmission of signals along the baseline had not been solved. The limits seemed to be in the region of baselines of not more than 10 to 50 km. Since equipment of sufficient sensitivity at much shorter wavelengths had not yet been produced it was not feasible to make any large change in the value of λ/d by conventional methods.

During 1950 we had many discussions on this issue at Jodrell Bank from which emerged Hanbury Brown's idea of a completely new type of interferometer in which it was unnecessary to maintain the phase of the signals in the separate aerials up to the moment of combination and hence no limit was set to the extent of the baselines. The signals from the two aerials at each end of the baseline are fed into two independent receivers. This system destroys all information about the absolute phase in either aerial of any single radio-frequency component. However, if ω ω' are different frequencies emitted by different points on the source in directions θ_1, θ_λ then the phase difference between the two beat frequencies at the output of a square law detector will be

$$\Delta\psi = \frac{l}{c}[\omega\theta_1 - \omega'\theta_\lambda]$$

where l is the projected length of the baseline normal to the source.

This phase difference $\Delta\psi$ is equal to the difference in the relative phase of the two radio-frequency components in each aerial and the information about the angular size of the source is preserved by the relative phase of corresponding beat frequencies in the detector outputs. In the practical interferometer which was developed these beat frequencies were multiplied in a correlator which gave a direct current output proportional to $\cos(\Delta\psi)$. It can be shown that this output is the sum of the outputs due to all the beat frequencies produced by all pairs of points in the source, and that it is proportional to the square of the correlation coefficient normally measured by the conventional interferometer of the Michelson type where the phase at the separate aerials is preserved.

The transference of these ideas into a practical system was a severe technical problem at that time. It was our good fortune that during the discussion of this problem a new research student came to Jodrell Bank in the autumn of 1950. R. C. Jennison's student career had been interrupted by the war. In 1942 he became an air crew navigator in the R.A.F. and was not demobilized until 1947. At the age of twenty-five he became an undergraduate in Manchester in order to finish his degree course. His interests and experience were such that he asked and was readily given permission to work with me at Jodrell Bank during the vacations. When he had obtained his degree in 1950 he came full time to Jodrell Bank to work for his Ph.D. degree. Jennison was exactly the man to execute Hanbury Brown's ideas. He possessed a practical knowledge of electronics and their application which were almost unique in a research student. With the help of a young Indian student, M. K. das Gupta, he built a practical version of the new interferometer with such speed that by the summer of 1951 he was able to measure the diameter of the Sun, thus demonstrating the correctness of the theoretical idea.

It was then necessary to develop this prototype instrument further for the measurements on the radio sources, and by the summer of 1952 he was ready to attempt the measurements at long baselines on the sources in Cygnus and Cassiopeia. Immediately success was achieved. Ironically both sources were found to be only slightly smaller than the limits set in the previous interferometer measurements. Cassiopeia was a nearly symmetrical radio source with a diameter of about 4 minutes of arc, but Cygnus was asymmetrical—2 minutes 19 seconds of arc in one direction but only 30 seconds of arc in another direction.

These were the first measured angular diameters of radio sources (as distinct from upper limits) to be published, but they were so close to the lower limits set by the conventional interferometers that it was no surprise for us to learn that a small extension of those methods had given similar results. In fact, while Jennison was making these measurements Hanbury Brown had gone to Australia for the meetings of U.R.S.I. (the International Scientific Radio Union). In August (1952) he wrote to me to say that he had seen B. Y. Mills who had succeeded in operating an interferometer over a baseline of 10 km using a radio link. 'Mills also says Taurus source i.e. Crab has got fairly large diameter i.e. >5 min. Apparently most sources are quite large. It seems we may have designed our interferometer to crack a nut which could have been more easily opened.' Simultaneously we learnt that in Cambridge, Graham Smith had obtained similar results by using a modification of the conventional interferometer.

Evidence rapidly accumulated that the radio sources were minutes of arc in diameter and not small fractions of a second of arc as with the common stars.* Technical improvements were making it possible to extend the baselines of the conventional phase correlation interferometers. Furthermore, the new intensity interferometer of Hanbury Brown depended essentially on the use of a square-law detector and was therefore

---

* The unfolding of the story of the angular diameter measurements will show that this generalized deduction about the angular diameter of the radio sources turned out to be quite erroneous.

relatively insensitive to weak signals and of little practical use unless the signal to noise ratio exceeded unity. There seemed to be no real argument for pursuing this complicated technique. It was used by Jennison* and das Gupta to evaluate the detailed structure of the radio source in Cygnus† and then, as a radio interferometer, fell into disuse.

This episode concerning Hanbury Brown and the development of the intensity interferometer transpired to have little consequence in the mainstream of development of radio astronomy. The measurements for which it was conceived were to be made by more straightforward developments of the phase correlation interferometers both at Jodrell Bank and elsewhere. It was, however, to have a memorable impact in quite unforeseen directions—in theoretical physics and in optical astronomy.

* R. C. Jennison was appointed to the Chair of Physical Electronics in the University of Kent at Canterbury in 1965.

† The dimensions of the Cygnus source quoted in this and subsequent chapters are those measured by Jennison and das Gupta in 1952. (R. C. Jennison and M. K. das Gupta, *Phil. Mag.*, Ser. 8, **1**, 65, 1956). In recent years more accurate measurements have substantially confirmed these dimensions. For example, the measurements on a wavelength of 11 cm by D. E. Hogg, G. H. MacDonald, R. G. Conway and C. M. Wade (*Astr. J.*, **74**, 1206, 1969) reveal the source as having two components separated by 123 seconds of arc along a position angle of 111·8° with each component having dimensions of about 30 seconds of arc along the major axis and about 15 to 20 seconds of arc across this axis.

3

The optical interferometer

During 1951 when Jennison was building the interferometer, Richard Twiss became a frequent visitor. He was then nominally a member of the Services Electronics Research Laboratory at Baldock in Hertfordshire. However, as a man of independent means and outlook, whatever his official duties may have involved he soon became immersed in the theoretical problems of the new interferometer. Over the next few years he concentrated his outstanding mathematical ability on the deep issues raised by this work.

In the beginning Hanbury Brown merely sought his advice and help on the theoretical problem of the radio interferometer. The results of this collaboration were published in July 1954 in a paper under the title of 'A new type of interferometer for use in Radio Astronomy'.* The last paragraph of this paper referred briefly to a subject which was being discussed constantly at Jodrell Bank.

The use of the 'Michelson' interferometer at radio wavelengths is a logical extension of optical practice and it is interesting to enquire whether the principle of the new type of interferometer can in turn be applied to visual astronomy, since in this way it might be possible to increase the resolving power and mitigate the effects of atmospheric turbulence. A preliminary examination of this question, which will be discussed in a later communication, suggests that the technique cannot be applied to optical wavelengths, and that it breaks down due to the limitations imposed by 'photon noise'.

The possibility of applying this new technique in optical astronomy opened dramatic possibilities. For a long time the arguments between Hanbury Brown and Twiss continued as to whether the transference of the system to the visual domain would be possible. Caution was engendered by an apparent conflict with fundamental physical theory. In order to use light waves the two radio telescopes would have to be replaced by two mirrors with photoelectric cells as detectors. The output of the photo-cells would then have to be brought to a correlator as in the radio case. Then, as a function of the separation of the mirrors, the

* R. Hanbury Brown and R. Q. Twiss, *Phil. Mag.*, Ser. 7, **45**, 663, 1954.

correlation between the fluctuations in the currents from the cells would be measured when the mirrors were directed at a star.

There were at least three major and perhaps decisive objections to this concept. First it was uncertain whether atmospheric problems might inhibit the correlation. However, similar fears as regards iono-spheric scintillation in the case of the radio sources had not been realized in practice. Before there could be any practical assessment of this possible hindrance in the optical case it was necessary to overcome two further objections of fundamental importance. It was essential to the operation of the system that the time of arrival of photons at the two photo-cathodes should be correlated when the light beams incident on the two mirrors were coherent. This effect had never been observed with light, and indeed some theorists we consulted maintained that the effect could not exist because it would be in violation of fundamental physical theory. Thirdly, even if the effect did exist it was not clear that the cor-relation would be fully preserved in the process of photoelectric emission.

These last two fundamental issues were the subject of a laboratory experiment at Jodrell Bank during 1955. The apparatus was constructed in a small room which had been built and used for a short time to house a spectrohelioscope. The light source was a mercury arc and a system of filters isolated the line at a wavelength of 4358 angstroms. The two coherent light beams were obtained by division at a half-silvered mirror. The separate light beams were fed from the half-silvered mirror to the two photomultiplier tubes along paths at right angles. The fluctuations in the output currents from the two cells were multiplied together in a correlator. The average value of the product recorded on the revolution counter of an integrating motor gave a measure of the correlation in the fluctuations.

Under the shadow of the rising structure of the telescope which was then causing us acute practical and financial problems, this experiment carried out in the small darkened room proved to be a famous land-mark in our history. The results were first published early in January of 1956 in *Nature*.* At the end of their account, the authors wrote: 'This ex-periment shows beyond question that the photons in two coherent beams of light are correlated, and that this correlation is preserved in the process of photoelectric emission. Furthermore, the quantitative results are in fair agreement with those predicted by classical electromagnetic wave theory and the correspondence principle.'

The publication of this paper caused a storm. People elsewhere carried out experiments to show that the interpretation of the results was erron-eous and several eminent theorists maintained that the conclusions were

* R. Hanbury Brown and R. Q. Twiss, *Nature*, **177**, 27, 1956.

in violation of fundamental quantum theory. Throughout all these arguments Hanbury Brown and Twiss stood their ground and eventually established a comprehensive theoretical rationalization of their results as between classical and quantum theory, in a series of papers published in 1957 and 1958 by the Royal Society.*

Meanwhile, having established these fundamental points in the 1955 laboratory experiment, the way was clear to a practical experiment on the stars. The immense importance of this was realized by all astronomers. The idea of applying interferometric principles to astronomy had been advanced in 1868 by Fizeau, but it was not until 1920 that Michelson and Pease carried out their famous measurement with a 20 ft interferometer mounted on the 100 inch telescope on Mount Wilson. With this they measured the diameter of Betelgeuse. Subsequently during the next ten years Pease made similar measurements on six more stars—all giants or super giants—whose diameters were in the range 0·047 to 0·020 seconds of arc. The practical difficulties in extending the baselines to as little as 50 ft were never overcome satisfactorily, partly because great accuracy (of the order of the wavelength of light) was needed in the optical paths, and partly because of the effects of atmospheric scintillation.

Thus, at the time of these experiments at Jodrell Bank, only about half a dozen of the giant stars had been the subject of practical diameter measurements. Our knowledge of the sizes of the stars had been derived largely from spectroscopic measurements of effective temperature, eclipsing binaries and, in a few cases, lunar occultations. Much uncertainty existed about the diameters of some of the hotter stars of type O, B. To measure these, and the nucleus of some of the Wolf-Rayet stars, baselines of more than a mile were believed to be necessary—entirely beyond the possibility of the Michelson type of interferometer.

It was against this background that Hanbury Brown and Twiss built a prototype version of this optical intensity interferometer to make a practical test of the system on Sirius. The choice of the star Sirius for the experiment was made because it was the only star bright enough to give a workable signal with the apparatus then available. Two ex-army searchlights on their mounts were borrowed to serve as the mirrors. They were placed several yards apart outside the window of the new control room. At that stage the room was almost empty of the control equipment for the telescope, and the recording equipment attached to the searchlights was placed in this room. The observations were made in two stages. In November and December 1955, and then with increased

* R. Hanbury Brown and R. Q. Twiss, *Proc. R. Soc. London*, Ser. A., **242**, 300, 1957; **243**, 291, 1957; **248**, 199, 1958; **248**, 222, 1958.

spacings during January to March 1956. The investigation was completely successful and the diameter of Sirius was measured as 0·0063 seconds of arc. The anxieties about the possible detrimental effects of atmospheric scintillation were unfounded and the first practical measurement of a stellar diameter for nearly thirty years had been achieved. The results were published in *Nature* in November 1956* while arguments about the interpretation of the laboratory experiment were still proceeding.

This brilliant and unexpected diversion of the radio interferometer technique was soon to have consequences which none of us foresaw. Naturally, having made the prototype observations Hanbury Brown proceeded to design, and get a grant from the Department of Scientific and Industrial Research to build, a much larger device with mirrors 23 ft in diameter, each composed of 250 small spherically curved, hexagonal shaped mirrors and moving on a circular railway track of 600 ft diameter. The need for clearer skies than those of Cheshire, the desire to measure stars of the Wolf-Rayet type visible in the Southern hemisphere, and the circumstance that Twiss had moved to Australia led eventually to the shipment of the new interferometer to New South Wales, where it was erected in a suitably remote spot at Narrabri. Hanbury Brown who had been appointed to a personal chair of radio astronomy in 1960 was given leave of absence in 1962 by the University of Manchester. Alas, the leave extended from one year to two, and then to a final break when he decided that in fairness to all concerned he ought to accept the offer of a chair in Sydney and remain there to use his new interferometer—which he did with brilliant success. To my great sorrow, our expectation and assumption that he would return to continue his work with the radio telescope never materialized and since few experimentalists of his calibre exist in the world he left an irreplaceable gap. For me it was an emotional and sad moment when in January 1964 I placed on record before the University Senate the account of the Ph.D. candidate who fourteen years later was leaving us as one of our distinguished Professors in the international field of astronomers.

* R. Hanbury Brown and R. Q. Twiss, *Nature*, **178**, 1046, 1956.

4

The development of long baseline interferometers at Jodrell Bank

The intensity interferometer had been developed for the specific purpose of measuring the angular diameter of those few radio sources which were strong enough to give sufficient signal strength through the square-law detector characteristics of the device. The initial measurements made by the system in 1952 showed that these sources were several minutes of arc in diameter. Simultaneously, the baseline of the phase correlation interferometer had been extended sufficiently in Cambridge and in Australia to resolve these sources. At that time therefore it seemed that nature had, after all, determined that in the radio part of the spectrum the discrete sources had an angular extent of minutes of arc, compared with the few thousandths of a second of arc of the common stars.

From these considerations it followed naturally that our interest in angular diameter measurements at Jodrell Bank turned to the developments of the phase correlation interferometer using the 218 ft transit telescope as one of the aerials. This would give us far more sensitivity than was available anywhere else at that time and should therefore bring many fainter sources into the scope of the measurements. Thus it happened that, as the concept of the intensity interferometer was in process of transformation to the optical domain, the line of development of the radio link phase correlation interferometer was initiated.

Over the years we have frequently been fortunate in adding to our staff exactly the right man at the critical time for new developments. Henry Palmer came to us at this precise moment in 1952. He was an Oxford man, having graduated in 1947 and getting his D.Phil. in 1952 for work on meteorology in the Clarendon laboratory. H.P. to B.L. 24 June 1952: 'Prof. Blackett has just told me that I may regard my appointment as Assistant Lecturer as definite. . . . I am looking forward very much to coming to Jodrell Bank, and I hope that I shall in fact find myself able to do some useful work.'* Palmer's interests were immediately directed to this development of the phase correlation interferometer. When Hanbury Brown left us ten years later Palmer had become the undisputed leader of that important section of our researches and continues in that position to this day. His hope that he might find himself able to do some useful work has materialized in researches of fundamental importance to the development of radio astronomy which, after 1957, monopolized nearly a quarter of the operational time of the telescope.

However, when Palmer arrived at Jodrell Bank work was only commencing on the foundations of that telescope and the possibility of using it for measurements of any kind seemed infinitely distant. Our important asset at that time, compared with the other groups who had already started measurements with phase correlation systems, was that we had a large radio telescope, which, although not steerable, could be used to study a strip of sky near the zenith. Beam shifting from the vertical position was made by the simple expedient of adjusting, by hand, the lengths of the hawsers securing the 120 ft feed tower.

This telescope had already been used by Hanbury Brown and his assistant Hazard on a wavelength of 1·89 m to survey the radio emission in the zenithal strip. They had measured the position of 23 localized sources and an obvious first task for the interferometer was to find out how large these sources were. A small broadside array was constructed on a wooden frame. The area of this mobile array was only 35 square metres, but in association with the transit telescope it was adequate to measure these sources. The technique was to place the mobile array at various spacings on an east–west line from the transit telescope and observe the fringe pattern as the rotation of the Earth swept the polar diagram of the combined aerials across the source—that is the telescope was used as a meridian transit instrument.

Two of the sources which were in the field of view were the extragalactic radio source in Cygnus and the nebulosity in Cassiopeia. For six others it was found that as the spacing between the aerials was increased, the amplitude of the fringes fell to zero at a spacing of about 50 wavelengths. The sources were therefore of large angular diameter, that is of the order of 1 to 3 degrees. At that moment there was still much speculation as to the nature of the localized sources. Only four had been identified with galactic objects and these were all filamentary nebulosities extending over a considerable area of sky (the Crab Nebula, Cassiopeia, Puppis, Gemini). This survey added a fifth, since Min-

* Subsequently Palmer was promoted Lecturer in 1955, Senior Lecturer in 1963, and Reader in 1967.

kowski soon identified one of these Jodrell sources with a nebulosity in Auriga.*

The indications at that time were that there might be two classes of radio source. One class seemed to be concentrated in the galactic plane, members of the galaxy and large diameter nebulosities of the type already discovered and probably supernova remnants like the Crab. In the other class the sources were more isotropically distributed, but notwithstanding the identification of the Cygnus source with a remote extragalactic object, a strong body of opinion still existed which maintained that these sources were largely galactic objects—probably unidentified star-like objects.

Hence the measurement of the angular diameter of the remaining sources detected by the transit telescope which lay several degrees from the galactic plane immediately presented itself as a problem needing urgent solution. There were five sources in this survey with galactic latitudes greater than ±5 deg and the angular diameter measurement of these was the target of the next investigation. In the earlier experiment the six large diameter sources had been resolved completely at baselines of about 50 wavelengths, whereas the five at high galactic latitudes showed little sign of resolution at a baseline of 500 wavelengths. The first of the new technical problems presented was therefore to extend this baseline. The wavelength of 1·89 m was near the working limit of the 218 ft transit telescope and hence it was necessary to increase the physical separation between the aerials. It had to be anticipated that a considerable increase over the 500 wavelengths (or nearly 1000 metres) separation might be necessary and it was not feasible for economic reasons to envisage a significant increase in a phase stable cable link between the two aerials.

The only practicable alternative at that time† was the development of a radio link to replace the cable. The essential function of the radio link is to convey information from the remote site to the home station in such a form that it can be used to produce interference fringes by multiplication with the signal from the home radio telescope. There were two major difficulties. Firstly, the phase relationship and the relative amplitudes of all components in the signal had to be preserved during the transmission. Secondly, the phase difference between the signal received at the remote station and the same signal as reconstructed at the home station must not vary by more than a fraction of a radian during

---

* The results of this investigation were published by R. Hanbury Brown, H. P. Palmer, and A. R. Thompson in *Nature*, **173**, 945, 1954.

† More than a decade later developments eliminating any linkage described in Ch. 6 became possible.

the observation of a source. The first requirement implies that the bandwidth of the radio link must be at least as broad as that of the signal. Although in observations of this type it is possible, in principle, to use narrow bandwidths in the receivers, in practice, bandwidths of at least 1 MHz are normally used since the signal–noise ratio varies as the square root of the bandwidth. Frequency allocations for radio astronomy of this bandwidth are exceedingly hard to obtain and are governed by national and international regulations. However, the development of microwave radio links for television purposes eventually eased this problem. They are limited to line of sight transmission and the problem of allocation is essentially a local national one. At that moment, though, radio links in the microwave region were not available and we had to construct our own link working on about 206 MHz. Many years later we found it possible and expedient to make use of commercial microwave equipment with bandwidths of a few megahertz.

The second requirement, that of phase stability, is harder to achieve. Essentially it implies that the local oscillator in the receiver at the remote site must be precisely reproducible at the home station. The solution was to drive the local oscillator in the remote receiver by means of a modulation on the microwave link. However, the properties of the link transmitter made it impossible to modulate at the required local oscillator frequency of 167·3 MHz—the limit of modulation frequency being 10 MHz. Frequency multiplication of about 10 MHz by 16 times carried too much danger of inherent instability. The solution involved a separate radio link on 175 MHz. The details of the rather complicated arrangements which were evolved to lock both remote and home local oscillators by means of this link are described in a paper which was published at a later date.* Although the link frequency was subsequently raised to 458 MHz, the system of control remains substantially unchanged to this day. Phase stabilities of better than 1 cycle per second in 10 minutes are achieved in this system.

Two other observational problems quickly presented themselves. The existing measurements had all been made with the aerials separated on an east–west baseline, so that the rotation of the Earth swept the lobe pattern over the region of the sky containing the radio source. Of course, this produces information only about the angular extent of the source along the line of transit. This would be of little consequence if the radio sources could be assumed to be spherical, like a star. But evidence already existed that this was not the case. Indeed, the measurements which Jennison and das Gupta were then making on the Cygnus radio source

* O. Elgaroy, D. Morris, and B. Rowson, *Mon. Not. R. astr. Soc.*, **124**, 395, 1962.

with the intensity interferometer showed that the radio emission extended for 2 minutes 19 seconds of arc in one direction (position angle 90°) but for only 30 seconds of arc in a direction at right angles, and furthermore that the source consisted of at least two prominent centres of emission. This was striking evidence of the need to measure the angular diameters in a number of directions across the source in turn, implying that the aerials must be separated in directions along any geographical baseline including north–south. Now the system had fixed aerials and the sources were observed only when in transit across the normal to the baseline of the aerials. The frequency of the interference pattern produced on the chart of the recorder as the Earth rotates is given by

$$f = \frac{1}{2\pi} \left[\omega_e . \frac{d}{\lambda} \cos \delta . \cos a \right] \qquad (1)$$

where ω_e (radians per second) is the rate of angular rotation of the Earth, λ the radio wavelength, d the distance between the aerials, δ the declination of the source, and a the angle between the direction of the baseline and the east–west line.

With the aerials fixed this frequency f depends only on the rotation of the Earth ω_e^*. If the baseline is east–west then $\cos a$ is unity. But as the baseline approaches north–south so $\cos a$ approaches zero and the frequency of the pattern becomes too low for satisfactory observation. Even with substantially east–west baselines another problem arises with the fringe frequency f at long baselines. In this case d may become a large number and hence the frequency of the pattern f becomes high. Ultimately f would become so high that the time constant of the receiver would have to be reduced. Since the signal–noise ratio is proportional to the square root of the time constant this would impede the observation of faint sources at long baselines.

Both these considerations pointed to the need to obtain some control over the fringe frequency f, other than that determined by the rotation of the Earth. The solution, in the form of the rotating-lobe interferometer, was devised by Hanbury Brown and Palmer, and constructed by them with the assistance of a research student A. R. Thompson*. The principle of this device is illustrated in Fig. 1. A and B are the two aerials separated by a distance d and connected to the multiplier and recording system either by cable or radio link. In one of the arms of the system there is a continuously variable phase shifter. If the radio source lies at an angle of θ_0 to the normal to the baseline between A and B and if a

* A. R. Thompson worked as a research student at Jodrell Bank from October 1952 until December 1955. Subsequently he emigrated and continued his radio astronomical work at Stanford, California.

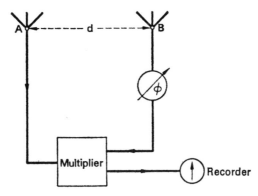

1. A simple form of interferometer.

phase shift of φ is introduced then the deflection of the recorder will be proportional to

$$P_s[G_A G_B]^{\frac{1}{2}} \cos \left[\frac{2\pi d}{\lambda} \sin \theta_0 + \varphi \right] \qquad (2)$$

where P_s is the flux from the radio source and G_A G_B are the power gains of the two aerials in the direction θ_0.

The response of the whole system to a radio source at any angle may then be represented by the polar diagram in Fig. 2, with lobes of alternate polarity. The angular position of the central lobe with respect to the normal is determined by the value of the phase shift φ.

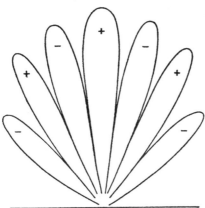

2. The effective polar diagram of the simple interferometer shown in Fig. 1.

The frequency of the fringe pattern with fixed aerials has been given in equation (1). If now the value of the phase shift φ is made to vary slowly with time at a rate ω_φ radians per second, then the lobes will appear to rotate although the envelope of their maxima remains stationary. The frequency of the pattern given by equation (1) is then modified and becomes

$$f_s = \frac{1}{2\pi}\left[\omega_e\,\frac{d}{\lambda}\cos\delta\cos a\pm\omega_\varphi\right] \tag{3}$$

The choice of the sign depends on the direction of rotation of the phase shifter and by controlling this and the speed it is possible to vary at will the value of the fringe speed f_s quite independently of the parameters ω_e, d, δ and a. That is, for baselines approaching north–south the speed of the pattern can be increased, or alternatively for large values of d the speed can be reduced.

The system developed to carry this principle into operation used a rotating magslip phase-shifter driven at an adjustable speed by a velodyne motor, and a phase-sensitive rectifier. The details were published

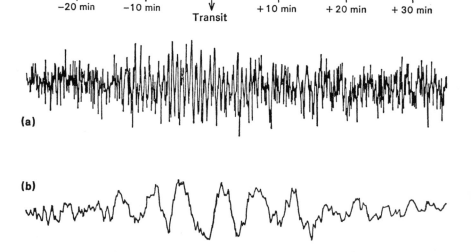

3. The use of a rotating lobe interferometer to decrease the frequency of an interference pattern.

(a)   Record of a weak source observed with a baseline of 630λ in a direction east and west, taken with the phase-shifter stationary. The output time constant is 4 seconds.

(b)   Record of the same source taken under identical conditions but with the phase-shifter rotating and decreasing the frequency of the pattern, and the output time constant increased to 30 seconds.

N.B.   For both these records the 218 ft paraboloid was used as one aerial of the interferometer. At a frequency of 158·2 MHz the beam is narrow and the interference pattern is therefore restricted to about 10 minutes on either side of transit.

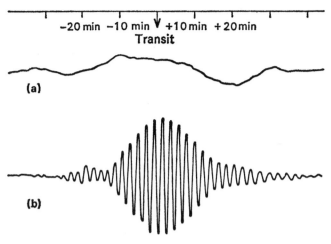

-20 min   -10 min   ▼   +10 min   +20 min
Transit

(a)

(b)

4. The use of a rotating lobe interferometer to increase the frequency of an interference pattern.
(a) A record of an intense source taken with a baseline 50λ in a direction north–south and with the phase-shifter at rest in an arbitrary position.
(b) A record taken under identical conditions but with the phase-shifter rotating and increasing the frequency of the interference pattern.

in 1955* with examples showing the effect of slowing down the fringe frequency f at long baselines (Fig. 3), and of speeding up the fringe frequency for observations on a north–south baseline (Fig. 4).

Measurements with this new system began in 1954. As before, the aerials were the 218 ft transit instrument and the small 36 square metre broadside array. It will be recalled that 6 of the sources lying in the galactic plane had been resolved with a baseline of 50 wavelengths (95 metres) indicating angular diameters of a degree or more, but that the 5 sources lying away from the galactic plane which were the object of this new investigation had shown no sign of resolution at 10 times that spacing. This new experiment began in June 1954 with the remote aerial at a spacing of 480 wavelengths (0·91 km), situated in a field near Bomish Lane to the north of Jodrell Bank—the field which was eventually to be turned into a visitors' car park. The measurements confirmed that even with a nearly north–south baseline the sources were unresolved at this spacing, indicating that they were less than a few minutes of arc in extent.

The progress of the experiment over the next three years then became a question of finding more and more remote sites on which to place the transportable aerial. All the time we were expecting that the next move

* R. Hanbury Brown, H. P. Palmer, and A. R. Thompson, *Phil. Mag.*, Ser. **7, 46,** 857, 1955.

would resolve all the sources and hence the spacings were carried out roughly in terms of successively doubling the baselines. Of course we depended heavily on the goodwill of the farmers in the vicinity who allowed the aerials and equipment in a small hut to be placed on their land. In March 1955 the site was at Smith's Green in Lower Withington (630 wavelengths, 1·2 km), in April 1955 at Ivy House Farm, Lower Withington (980 wavelengths, 1·84 km), in June 1955 at New Platt Lane, Goostrey (2060 wavelengths, 3·92 km), and in August/September 1955 at Timbersbrook near Congleton (6700 wavelengths, 12·8 km).

At this stage of the programme 2 of the sources had been resolved, but 3 remained, showing no sign of a change in the fringe amplitude at the longest baseline of 6700 wavelengths, implying that their angular sizes must be less than 24 seconds of arc. At the beginning of 1956 the equipment was operated at an even more remote site near the Cat and Fiddle Inn in the Peak District (10 600 wavelengths, 20 km). The 3 sources remained unresolved showing that their angular diameters were less than 12 seconds of arc. These were by far the smallest diameter sources measured at that time. In their short report* after the conclusion of the experiment, Palmer and his colleagues pointed out that the brightness temperature of the three sources must be at least that of the remote source in Cygnus (10^8 K at that frequency) and this suggested that the three sources 'may be objects of the same type'.

The identification of a remote galaxy—the impact of the Mark I telescope

When I reflect on the remarkable consequences which were to follow from these measurements I am surprised to find that there was a hiatus in this work. The next published note from Jodrell Bank did not appear for three years. Was this an error of judgement or was the interregnum forced on us in some way? Palmer was, by this time, deeply involved as the leader of this research group and recently I asked him if he could recall why I had not insisted that they should again double the baselines in 1956 and so still further reduce the limits of the angular diameters.

As so often happens in these matters the answer is complex. Palmer reminded me, for example, that in 1956 we were so surprised to find that the radio sources were less than 12 seconds of arc in diameter that we considered it expedient to repeat the measurements at the Cat and Fiddle. The apparatus was installed again at the remote site but a winter had intervened and when we attempted to operate in 1957 we found that the radio link frequency near 200 MHz had in the intervening period been

* D. Morris, H. P. Palmer, and A. R. Thompson, *Observatory*, **77**, 103, 1957.

allocated to I.T.V. so that our link was completely jammed. A new allocation could only be made in the microwave band and the construction of a suitable link was a lengthy affair for us. Furthermore, at that period everyone was under pressure to have new apparatus ready for the Mk I telescope which was nearing completion—and this included the small interferometer group.

Even so, I have no doubt that these handicaps would have been surmounted if the real significance of further measurements had been appreciated. In 1958 I was present at two important meetings: in the spring in Brussels at the Solvay Conference on 'The Structure and Evolution of the Universe', and later, in August, at a full-scale Radio Astronomy Conference in Paris. At the Solvay Conference I spoke on the subject of 'Radio astronomical observations which may give information on the structure of the universe'.* The data which I presented there were not added to significantly by the much wider circle of astronomers who were present later in Paris. I said that in the local galaxy 3 supernovae, 5 peculiar gaseous nebulosities (probably supernovae), and 15 emission nebulae of ionized hydrogen near hot stars had been identified as radio sources: 'Apart from these few cases the attempt to associate the bulk of the radio sources with galactic objects, either on an individual or statistical basis has failed'. I then proceeded to comment on the 'known extragalactic radio objects and then . . . the problem of the unidentified radio sources in so far as they may give information about the structure of the universe'.

At that time the Cambridge and Sydney counts of radio sources numbered well over 2000, and yet the list of the identifications which I gave amounted only to 16 normal galaxies and 7 of the abnormal extragalactic objects, discovered by Baade and Minkowski. These, headed by the radio source in Cygnus, believed then to be colliding galaxies, had a ratio of radio to light output of the order of a million times greater than the normal galaxies. In addition to Cygnus, the extragalactic objects identified with radio sources were the strange objects in Perseus (NGC 1275), Centaurus, Fornax, Hercules, Hydra, and M87 in Virgo. There were also a few probable cases of radio emission from clusters of galaxies. The arguments for the extragalactic nature of the majority of the radio sources rested on their isotropic distribution and the implied evidence from the distribution of number and intensity in the Cambridge and Sydney counts: '. . . the positive evidence in favour [of their extragalactic nature] depends on the association of 4 sources with interacting galaxies, together with one rather uncertain case and two sources with peculiar nebulae having a high ratio of radio-optical

* A. C. B. Lovell, *La Structure et l'evolution de l'univers* (Brussels, 1958).

emission.' As one of the important requirements, I mentioned 'the measurement of the angular diameter and intensities of the individual sources [since] . . . the determination of the brightness temperature may be a significant factor in assessing the class and distance of the object'.

Unfortunately we were preoccupied and distraught with the practical, political, and financial problems of the Mk I telescope. We were not seized with the immense significance of these measurements until Minkowski announced in 1960 that he had identified one of these three sources whose diameters were less than 12 seconds of arc. Naturally, Minkowski had been acquainted with the results of the angular diameter measurements and had realized that because of the high brightness temperature, similar to that of the Cygnus source, these radio sources might well be similar objects at greater distances.

The object on which Minkowski concentrated his attention was 3C 295 (in the Cambridge catalogue nomenclature). With a diameter of less than 12 seconds of arc, that is 10 times smaller in angular extent than that of the source in Cygnus according to the measurements then available from Jennison's work, and 70 times fainter, it seemed clear that this might be a distant analogue of Cygnus A. The first attempt which Minkowski made to photograph the object using the 200 inch telescope was unsuccessful because the position of the radio source was not known with sufficient accuracy. However, in 1959 Ryle and his colleagues in Cambridge obtained a more precise position, and further confirmation of the position in right ascension by Bolton at Owens Valley encouraged Minkowski to repeat the attempt at identification.

Minkowski announced in 1960* that the photographs taken with the 200 inch showed the presence of a distant cluster of galaxies surrounding the position of the radio source. There were 60 galaxies in the range of visual magnitude 21, to the limit of the 200 inch in an area of 3 minutes of arc in diameter in the centre of the plate. One of the brighter galaxies in this group with visual magnitude 20·9 was precisely in the position of the radio source 3C 295. Further, since the spectrum of that galaxy resembled that of Cygnus A in showing an unusually strong emission line, Minkowski had little hesitation in concluding that this was the object responsible for the radio emission.

With the nebular spectrograph at the prime focus of the 200 inch telescope Minkowski obtained two spectrograms of the galaxy, one with an exposure of 4·5 hours and another with 9 hours exposure. The emission line was identified as that of $[O_{II}]$ — the same strong emission line visible in the Cygnus source. The unshifted wavelength of this line is 3727·7 angstroms, on the spectrogram the measured wavelength was

* R. Minkowski, *Astrophys. J.*, **132**, 908, 1960.

5447·8 angstroms, indicating that the galaxy was showing a redshift of $\Delta\lambda/\lambda_0 = 0.4614$. Further photometry indicated a redshift of 0·44 for the whole cluster and hence it was clear that the galaxy with the strong emission line was a member of this cluster. In this manner Minkowski identified by far the most distant object known in the universe at that time. The measured redshift of $\Delta\lambda/\lambda_0 = 0.46$ implied a recessional velocity of more than 40 per cent of the velocity of light and a distance of about 4500 million light years.

This clear demonstration of the belief that the small-diameter radio sources might be objects analogous to Cygnus but at much greater distances was instantly realized to be of immense significance in cosmology. The predictions of the various evolutionary theories and of the steady-state began to diverge significantly in the regions of space and time to which the telescopes had now penetrated by the identification of this 3C 295 radio source. Furthermore, at the Paris Symposium in August 1958, Fred Hoyle had spoken on 'The relation of radio astronomy to cosmology'* As a possibility for a decisive test between his steady-state and the Einstein–de Sitter type of evolutionary cosmology he had drawn attention to the peculiar difference between the apparent angular diameter of a source of standard size on the two theories. For a source of diameter D, and with the Hubble constant H, the apparent angular diameter decreases as the redshift increases on the steady-state theory, tending asymptotically to the value DH/c. But on the Einstein–de Sitter cosmology the apparent angular diameter decreases as the redshift increases but reaches a minimum value with the apparent diameter of $(\frac{3}{2})^3 DH/c$ at a redshift $\Delta\lambda/\lambda_0 = \frac{5}{4}$. At greater redshifts the apparent angular diameter then increases and ultimately as $\Delta\lambda/\lambda_0$ approaches infinity the object would fill the whole sky.

Hoyle showed the diagram reproduced in Fig. 5 giving the results of these calculations of the apparent diameter $\Delta\theta$ of a source of absolute diameter D, plotted against redshift. Taking Cygnus as a typical source for which $\Delta\lambda/\lambda_0 = \frac{1}{18}$ and $\Delta\theta = 80$ seconds of arc, then the arbitrary scale factor in Fig. 5 is determined. For a Hubble constant H of 4×10^{17} seconds^{-1}, then the asymptotic value for the steady-state theory becomes 4 seconds of arc and the minimum for the Einstein–de Sitter model is 15 seconds of arc. The general property of a minimum angular diameter is common to all the forms of relativistic cosmology, although the precise value of the minimum depends on the assumption made about the various parameters. Hoyle said that these considerations seemed to give the best immediate hope of subjecting cosmological theory to obser-

* F. Hoyle, in *Paris Symposium on Radio Astronomy 1958*, Ed. R. N. Bracewell (Stanford, 1959), p. 529.

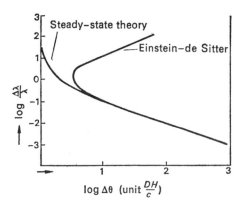

5. Apparent diameter $\Delta\theta$ of a source of absolute
diameter D, plotted against redshift.

vational test. Since the baselines used in these angular diameter measurements seemed to be close to those needed to make this observational
test, we had a powerful incentive to extend this work to somewhat
longer baselines.

The apparatus which had been developed for these interferometer
measurements, using the new Mk I telescope at Jodrell Bank as the home
station, was ready. The records of telescope use show that between 26
November 1958 and 18 October 1960, more than 3000 hours, or 37 per
cent of the telescope time, was used in this work. By the time these
measurements had been completed in 1961 about 4000 hours had been
spent on them. The previous measurements were made with the transit
telescope and hence the observations were limited to the zenithal strip of
sky; now the Mk I could be directed to any elevation. New aerials for
the remote site had been constructed, so that their elevation could be
varied. These consisted of three cylindrical paraboloids, each with a
length of 30 ft and width of 25 ft. They were formed by aluminium
wires stretched horizontally on tubular aluminium frames. The three
cylindrical paraboloids were placed end to end and, with a working
frequency of 158·6 MHz, the beam-width was 4 deg × 16 deg, and the
effective collecting area 100 m². The aerials were fixed in azimuth but
could be moved by hand in elevation and set at 5 deg intervals. Thus,
with the Mk I telescope set at the same azimuth and elevation as the
cylindrical aerials, the rotation of the Earth scanned the sky in right
ascension and by changing the pre-set elevations the whole sky between
declinations −20° and +10° was eventually observed.

The electronic equipment was an improved version of that used
earlier, with the transit aerial and small broadside array, using the rotating lobe principle. The reasons for desiring control of the fringe fre-

quency have already been described. In the earlier work this had been achieved by rotating the phase of an audio frequency signal using a phase resolver magslip. For short baselines it had been found convenient to rotate the magslip by coupling it to a velodyne controlled by a calibrated d.c. voltage. At long baselines the system was not satisfactory. For example at 32 000λ the fringe frequency for a source on the celestial equator is 140 cycles per sidereal minute and this had to be reduced to one cycle in 3 minutes for satisfactory recording on the chart—a reduction of over 8000 times. Since it was desirable to control the fringe frequency on the chart to $\pm$ 20 per cent, the rate of rotation of the magslip had to be controlled to one part in 2000. Furthermore, since the frequency depends on the cosine of the declination of the source, the rate had to be variable by 3 to 1 to cover all the declinations observed. In order to do this Rowson* devised a system in which a wheel with 60 slots was mounted on the shaft of the velodyne. This wheel interrupted a beam of light falling on a photo-transistor. The signal from the photo-transistor was counted down by a variable counter and the output compared with the pulses from a pendulum-controlled sidereal clock. The error signal was then applied to the velodyne. This contraption with gearing and mountings made of Meccano was a familiar feature of the interferometer laboratory for many years, contrasting strangely with the polished elegance of commercial products. It was cheap and effective, working reliably and enabling the speed of the velodyne to be varied by 100 to 1 in steps never more than a two-thousandth of the maximum speed and with an accuracy governed only by that of the sidereal clock.

The last series of measurements which set the limits of 12 seconds of arc had been carried out with the remote aerial at the Cat and Fiddle, 20 km distant—a baseline of 10 600 wavelengths. At 1690 ft there was a clear line of sight to Jodrell Bank. Further extensions of the baseline in that direction would have engulfed the remote aerial in the hills and dales of the Peak District and there was no obvious chance of getting a line of sight in that direction from Jodrell Bank at the much greater distances which we were now seeking. Furthermore, the existing link frequency had been jammed by I.T.V. as already described and it was necessary to move to the microwave region which demanded an even more stringent line of sight condition. Attention was therefore turned to the west and after the system had worked locally in Goostrey (2200λ, 4·2 km), and Northwich (9700λ, 18·4 km) a site was found at Holywell

---

* Dr. Barrie Rowson came to Jodrell Bank as a research student in 1954 after graduating in the University of Manchester. In 1957 he was appointed to a Leverhulme Research Fellowship and in 1962 to a Lectureship in Radio Astronomy.

in North Wales at a distance of 60·3 km which gave a baseline of 32 000 λ. After operating at that site during the summer of 1960 it was clear that even longer baselines must be sought.

At that time we had an outstation on the Lleyn peninsular near Abersoch. This was concerned with our meteor researches. A Schmidt camera-telescope was installed there for simultaneous photographic and radio echo studies of meteors. The remote radio equipment worked on a low frequency in the 30 MHz region and linkage, using a frequency of 2·8 MHz, with the home based transmitter-receiver at Jodrell Bank was no problem. The site was possible for the interferometer and naturally my interest was aroused by the possibility of using Snowdon as a site for the repeater station. I argued with Palmer that there must be a line of sight from the top of Snowdon to Jodrell or at least to one of the nearby hills. However, Palmer did not seem to like the idea of working near the top of Snowdon and he did not investigate the possibility with much alacrity, claiming that several hundred feet of Moel Famau must be in the line of sight. The situation was saved by the fact that during the course of these arguments Hanbury Brown drove home one day from a visit to a northern university on the road which runs close to the B.B.C. station at Holme Moss in the Peak District. The sight of microwave aerials on those very tall masts was a clear indication to him to add another. Negotiations with the B.B.C. proceeded without difficulty. There was a clear line of sight to Jodrell Bank and in the other direction to a disused airfield in Lincolnshire which we found for the remote aerials and equipment.* So the B.B.C. TV mast was soon in use as our microwave repeater station, and the measurements at a baseline of 61 000 λ over a distance of 115·4 km were made in 1961.

The details of the solution of these problems and other associated ones have been described by Rowson, who was responsible for many of the technical solutions, and two young students then working with him in Palmer's group D. Morris and O. Elgaroy a visitor from Norway.† Their paper also contains the remarkable record reproduced in Fig. 6 of the radio source 3C 48, showing that this object was not resolved even at the longest baseline achieved with this system.

Between 1958 and 1961 this system was operated at the four different sites and the angular sizes of 384 radio sources were measured. By the time the measurements at the second of the sites (9200 λ) had been completed most of these sources had been resolved, although the results indicated that the structure must be complex in many cases. However,

* In fact it was most active. Only when the mess bills were tendered did we realize that the airfield had been converted to a Thor ballistic rocket base.

† O. Elgaroy, D. Morris, B. Rowson, *Mon. Not. R. astr. Soc.*, **124**, 395, 1962.

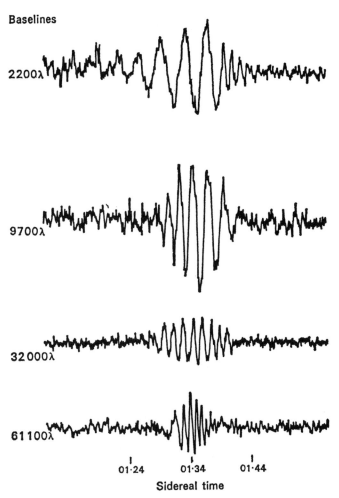

Baselines

2200λ

9700λ

32 000λ

61 100λ

01·24   01·34   01·44

Sidereal time

6. Examples of the large fringe patterns observed during the transits of the source 3C 48 at four long baselines.

the surprising feature was that about 10 per cent of the sources showed no signs of resolution, indicating that they must have angular diameters less than 8 seconds of arc. In 1960 the baseline was extended to the third site at Holywell and in November of that year Palmer and his collaborators published a short note which gave two important items of news.* Firstly, the source 3C 295 in Boötes which had been identified by Minkowski was resolved and its angular diameter which had previously been given as less than 12 seconds of arc was now measured to be 4½ seconds of arc. Secondly, that out of 91 sources measured at that base-

* L. R. Allen, H. P. Palmer, and B. Rowson, *Nature*, **188**, 731, 1960.

line at least 7 were still unresolved showing that their angular diameters must be less than 3 seconds of arc. The extremely high brightness temperatures thereby indicated supported the belief that these radio sources were remote objects in the universe perhaps even more distant than the one in Boötes—$4\frac{1}{2}$ thousand million light years. Already these results had produced the evidence that sources existed smaller than the 4 seconds of arc limit calculated by Hoyle as the minimum value to be expected for a Cygnus-like source in an Einstein–de Sitter model of the universe.

5

The discovery of quasars

It seemed evident that the identity of sources of such small angular diameter might well be of cardinal significance to the cosmological problem and that the objects in question formed a sufficiently homogeneous group to justify the belief that the small angular diameter sources were those at the greatest distances from Earth. The great importance of attempting an optical identification of these five sources was evident to all, but the existing positional measurements given in the Cambridge catalogue were not sufficiently precise. Fortunately, T. A. Matthews, at Owens Valley in California, and his colleagues were able to use the interferometer of two 90 ft steerable paraboloids to measure the right ascension and declination of one of these sources, 3C 48 with six times greater precision than that given in the Cambridge catalogue. The attempt at optical identification with the 200 inch Palomar telescope then became possible.

In New York City, between 28 and 31 December 1960, the American Astronomical Society held its 107th meeting. It was there that Allan Sandage announced that on plates taken with the 200 inch telescope on 26 September 1960 he had found in the position of 3C 48 a 16th magnitude star of a strange blue colour and with a faint wisp. The published note of his 'unscheduled paper' in *Sky and Telescope** contains the statement 'Since the distance of 3C 48 is unknown, there is a remote possibility that it may be a very distant galaxy of stars; but there is general agreement among the astronomers concerned that it is a relatively nearby star with most peculiar properties'. Shortly after the identification of 3C 48 Sandage made similar identifications of two of the other small-diameter sources 3C 196 and 3C 286 with star-like objects of a peculiar blue colour.

During the course of the next year 3C 48 was under regular observation and the photometric records showed that its light output was not constant but varied by at least 0·4 magnitude during that time. The concept that a galaxy of several thousand million stars distributed over

* *Sky Telesc.*, **21**, 148, 1961.

a huge volume of space could show such variations in brightness seemed so ludicrous that it could be discarded, and so the stellar nature of 3C 48 and of the other two objects seemed to have definite proof. So, for more than two years the astronomical world was startled by this discovery, that powerful small-diameter radio sources which had been believed to be remote objects in the universe, were apparently relatively nearby stellar objects in the Milky Way. Although there seemed to be this circumstantial evidence for the stellar nature of these small-diameter objects, their spectra could not be properly interpreted. Several strong broad emission lines were prominent but these could not be identified with any plausible combination of redshifted emission lines.

The observations which gave the key to the remarkable nature of these objects were made in 1962. They owe their initiation to another strange chain of circumstances involving a young man from Jodrell Bank. Cyril Hazard, who came from Cumberland, studied physics as an undergraduate in Manchester and graduated in 1949. He was in some of the classes to which I lectured and had been interested in the early developments at Jodrell Bank. I was therefore pleased to learn from Blackett that he was willing to allocate one of the department's D.S.I.R. maintenance grants to Hazard who began his research career with us early in October of that year. This was a few weeks after the arrival of Hanbury Brown and it required little consideration to place the new student under his care. This combination of Hanbury Brown and Hazard working with the 218 ft transit telescope yielded the harvest of results on the Andromeda galaxy already mentioned. Three years elapsed and with it Hazard's grant and also his temporary screening from military service. B.L. to the Recruiting Board: '... unfortunately the mass of data which has accumulated is not yet fully analysed and it is a matter of considerable urgency that this analysis should be completed before Mr. Hazard leaves Jodrell Bank since he alone is acquainted with many of the details and conditions under which these records were obtained.' After many months—B.L. to Hazard (who had gone home to Cumberland) 26 October 1952: 'I saw Mr. —— who deals with the deferments. ... I think you can safely assume that you will not be called up for the next 6 months or so. It is unlikely that you will obtain any definite statement from the authorities so I should not worry them any more.'

In the end we managed to keep Hazard for long enough to enable him to complete the analysis and write his thesis. Eventually he left us in the late summer of 1953 to join the Royal Naval Scientific Service where he satisfied the conditions of his call-up for military service. We had been much impressed by Hazard and so two years later in September

1955 I wrote to him: 'There is a high probability that we may have a vacancy as a temporary lecturer in radio astronomy for two years from this September. Both Hanbury Brown and I would be very glad if you were able to consider the position should the offer materialize in the near future.'* Hazard was only too glad of the chance to return and after the completion of the necessary formalities he was at Jodrell Bank once more in December 1955. He started work again with Hanbury Brown and began preparations for using the Mk I telescope. Just as he was beginning work with this his appointment expired but we found a vacancy in a University Research Fellowship.

With the new telescope Hanbury Brown and Hazard extended their observations of the radio emission from normal galaxies. One day Hazard asked me to read a memorandum which he had written. At that time few of the radio sources had been identified optically, the major problem being that the positions of the radio sources were not known with sufficient accuracy—the uncertainties being several minutes of arc. Other workers had already pointed out the possibility that the occultation of a radio source by the Moon might be observed and the technique had been used elsewhere in a few cases to study some intense sources of large angular diameter. Now, for the first time, we had available a very large steerable radio telescope and Hazard pointed out that if the telescope was made to track the Moon continuously then, if a radio source was occulted, the measurements of the time of disappearance and reappearance of the radio source would give the position of the source with high precision.

Naturally, I told him to get on with the work as quickly as possible and on 8 December 1960, a good opportunity occurred to test the feasibility of the system. Calculations showed that a radio source, 3C 212 in the Cambridge catalogue, would be occulted by the Moon between 8 a.m. and 9 a.m. Early in the morning the telescope was set in automatic motion to track the centre of the Moon's disc. At $08^h 03^m 52 \cdot 5^s$ a sharp fall in the signal strength was observed as the Moon covered the source, and a sharp rise when the source emerged at $08^h 40^m 49^s$. There were, of course, some uncertainties in these times because of the error in estimating the exact time of the immersion and emersion of the source; nevertheless when all allowances had been made Hazard was able to give the position of the source to $\pm 0 \cdot 3$ s in right ascension and ± 2 seconds of arc in declination—improvements of nearly 20 times in right ascension and 240 times in declination over the best previously known position. As Hazard wrote in his published note of this

* The temporary vacancy arose because Dr. Stanley Evans was given leave of absence to take part in the Royal Society's Expedition to Antarctica.

work* 'the measurements reported here probably represent the most accurate determination yet made of the position of a radio source'.

The radio source was occulted again on 31 January 1961 and although high winds put the telescope out of action for some of the time, Hazard made further observations to supplement the December measurements. His complete account of this work and of the potentialities of the method of lunar occultations using a large steerable telescope was not submitted for publication for another two years,† but by that time he had left us. Like many other young men he began to feel that he wanted to take the opportunities which scientific research offered to see more of the world.

It was then a simple matter for any competent young British scientist to find lucrative posts overseas and the consequent loss to British science has been a matter of national concern. In the decade to 1970 we lost between 40 and 50 per cent of the students who had taken higher degrees at Jodrell Bank in this way. The majority went to Australia or North America. Hazard sailed for Australia in February 1961 to take up an appointment as a Lecturer in Physics at Sydney University. However, a little over a year later, he wrote to me that he might not want to stay there for very long and there followed a good deal of correspondence, with Hazard trying to decide if he was interested in returning to us at Jodrell.‡

On 9 February 1963, I received a handwritten letter from him which I treasure as a classic of British understatement and phlegm in the face of great discoveries. Most of the letter complained about the temperature of 104° at Narrabri where he was carrying out some work with Hanbury Brown, and about the problem of his house and cost of moving his family. Then came the following paragraph about the work he had been doing with the steerable telescope at Parkes using the technique of lunar occultation:

The work I have done at Parkes has turned out very well. 3C 273 has now been identified with a 13th magnitude star! We noticed this star some months ago on a 200″ plate, it is a peculiar object with a jet protruding at an angle of 45° which is the inclination of the radio source. The source is double and the separation of the components approx. equal to the length of the jet (∼20″). A very intriguing identification.§

* C. Hazard, *Nature*, **191**, 58, 1961.

† C. Hazard, *Mon. Not. R. astr. Soc.*, **124**, 343, 1962.

‡ Subsequently he moved to Cornell and then back to England as a member of Fred Hoyle's Institute of Theoretical Astronomy in Cambridge.

§ In this letter Hazard uses the abbreviation ″ for two different purposes. 200″ refers to the 200 inch Palomar telescope. 20″ refers to an angular extent of seconds of arc.

Hazard and his colleagues published the details of these measurements in *Nature* on 16 March.* Simultaneously, Maarten Schmidt of Palomar published his comment on the identification:† 'The close correlation between the radio structure and the star with the jet is suggestive and intriguing. . . . The only explanation found for the spectrum involves a considerable redshift. A redshift of $\Delta\lambda/\lambda_0$ of 0·158 allows identification of four emission bands as Balmer lines.' After giving further evidence that the spectral characteristics supported the interpretation of a redshift of this magnitude Schmidt wrote that the unprecedented identification of a spectrum of an apparently stellar object in terms of a large redshift suggested two explanations—the object was either a star with a large gravitational redshift (but its radius would have to be only 10 km) or the nuclear region of a galaxy with a cosmological redshift of 0·158, corresponding to an apparent velocity of 47 400 km s^{-1}. However, this would imply a distance of 500 megaparsecs, and that the nuclear region would have a diameter of less than 1 kiloparsec. Further this explanation would imply that the nuclear region was 100 times brighter optically than the galaxies previously identified as radio sources, and that the energy being radiated in the optical region would be about 10^{59} erg. In spite of these remarkable consequences Schmidt concluded: 'At the present time, however, the explanation in terms of an extragalactic origin seems most direct and least objectionable.'

The success of Schmidt in identifying the spectrum of 3C 273 as Balmer lines with a large redshift led Jesse Greenstein and T. A. Matthews to consider again the case of the spectrum of the starlike object 3C 48 which had remained an enigma since the identification was announced by Sandage two years earlier. Previously it had not been found possible to interpret the spectrum in terms of a redshifted Balmer series, but now the possibility of very large redshifts of the magnitude found by Schmidt for 3C 273 was re-explored. They found that the observed spectrum could be made to fit a redshifted Balmer series if $\Delta\lambda/\lambda_0$ was 0·367—an apparent velocity of 110 200 km s^{-1}. This astonishing conclusion was published in the same issue of *Nature*.‡ Only rarely can a scientific journal have published a sequence of results of such enormous importance to astronomy. The enigmatic objects which had been believed for over two years to be stars in the galaxy were, on the contrary, found to be amongst the most distant objects known in the universe. Greenstein and Matthews wrote in their paper that 'so large a redshift,

* C. Hazard, M. B. Mackey, and A. J. Shimmins, *Nature*, **197**, 1037, 1963.
† M. Schmidt, *Nature*, **197**, 1040, 1963.
‡ J. L. Greenstein and T. A. Matthews, *Nature*, **197**, 1041, 1963.

second only to that of the intense radio source 3C 295, will have important implications in cosmological speculation. ... the distance of 3C 48, interpreted as the central core of an explosion in a very abnormal galaxy, may be estimated as $1 \cdot 1 \times 10^9$ parsecs.' As in the case of 3C 273 very large energies were computed, further 3C 48 radiates 50 times more powerfully in the optical region than other intense radio galaxies.

Two months later Harlan Smith and Dorrit Hoffleit announced in *Nature** that they had investigated the photographic history of 3C 273, 3C 48 and some of the other starlike objects. In the case of 3C 273 they were able to find the object on plates in the Harvard collection dating back to 1887. They found that the object appeared to be fluctuating in brightness with a period of about ten years and that there were occasions when the object increased in brightness by more than half a magnitude. This variability is equivalent to switching on and off the light of about 100 000 million Suns within short time periods and deepened the mystery as to the mechanism by which these newly discovered objects generated their enormous energies.

By the end of the year nine of the radio sources of small angular diameter had been identified with similar starlike objects and quite soon more of the spectra had been shown to be the Balmer series redshifted by very high amounts. Schmidt and Matthews† found that the spectrum of 3C 47 indicated a redshift of 0·425—which implied a distance of the order of 1275 Mpc. 3C 147 showed the largest redshift known, 0·545, implying a distance of 1635 Mpc.

In the following years more and more of these peculiar starlike objects were identified at even greater redshifts. Initially they were referred to as quasi-stellar radio sources or quasi-stellar objects. The name quasars which seems first to have been suggested by Hong-Yee Chiu of the Goddard Institute for Space Studies, did not find acceptance in the beginning but has now come into general use. Today we believe that the quasars must constitute a considerable proportion of the radio sources which are delineated by the large radio telescopes and that they may well provide the key to our understanding of the fundamental evolutionary processes in the universe.

The initial steps which led to this amazing sequence of events resulted from the angular diameter measurements made in 1960 between Holywell in North Wales and Jodrell Bank as described in Chapter 4. These measurements had shown that 7 of the 91 sources studied were unresolved at these baselines, implying that their angular diameters

* H. J. Smith and D. Hoffleit, *Nature*, **198**, 650, 1963.
† M. Schmidt and T. A. Matthews, *Astrophys. J.*, **139**, 781, 1964.

must be less than 3 seconds of arc. As the drama of the optical identification of these sources was unfolding, Palmer and his colleagues made further attempts to resolve these sources by doubling the baseline again. They moved the remote aerial from Holywell to the airfield in Lincolnshire and brought the signals back to Jodrell Bank via the repeater station on the TV mast at Holme Moss. During 1961 they continued their measurements over this baseline of 61 100 λ with a resolving power of about 1 second of arc.

At the previous baselines the remote aerials were set at a given elevation and left for twenty-fours. Now the remote site was manned continuously so that the elevation of the aerials could be varied several times per day in order to concentrate the measurements on the sources which were still unresolved over the 32 000 λ baseline. Fortunately for these remote station operators the measurements had to be made in the summer because the B.B.C. would not allow us to keep the repeater equipment on their mast at Holme Moss during the winter gales. Altogether in these series of measurements 384 sources were investigated. At the longest baseline of 61 100 λ there were 187 still remaining to be investigated.* Of these, 165 did not yield any fringes, showing that their angular diameters were greater than 1·5 seconds of arc. Of the remaining 22, five were estimated to be less than 0·8 seconds of arc in diameter.

Amongst these was the radio source 3C 48, which at that moment had been identified as the strange blue object, the history of which has already been related. The results of the analysis of this long series of measurements were published in 1962† before the recognition of the true nature of the blue objects. One of the important conclusions was that nearly half of the sources must have a complex structure, that is, at least two main radiating centres like that of the Cygnus radio source. On the question of the use of the size measurements for distinguishing between cosmological models, which was the major stimulation for the work, no final conclusions could be reached. There were 10 sources which had an angular size less than the asymptotic limit of 1·7 seconds of arc for objects of size 25 kpc in the steady-state model, and there were 22 sources with a diameter less than the minimum value of 5 seconds of arc in an Einstein–de Sitter model. The authors concluded:

* The observations at Holywell could not be made on as many sources as originally intended because the ending of the seamen's strike led to the resumption of the use of the radio telephones on the Mersey which jammed the equipment.

† L. R. Allen, R. Hanbury Brown, and H. P. Palmer, *Mon. Not. R. astr. Soc.*, **125**, 57, 1962.

If there were no dispersion in the linear size of radio sources, or if we knew this dispersion precisely, then it would be possible to discriminate between certain cosmological models on the basis of the data about angular size. In particular we might be able to rule out a range of models simply by observing the apparent cut-off in the angular size distribution for small diameters. However, this cannot yet be done because of the uncertainty about the dispersion in linear sizes, and also because we cannot be sure that our source population does not include some small diameter sources of a radically different type.

The tracking radio interferometer

The angular size of a radio source measured in these experiments was at the position angle determined by the geographical direction of the baseline between Jodrell Bank and the remote site. The evidence quoted above, coupled with the previous measurements on the Cygnus source, showed that many of the radio sources were far from having any circular symmetry and indeed that a considerable fraction must have more than one centre of emission. The answer about the 'size' of the source determined in this work was therefore treated with considerable caution. The Cygnus source had been found to be double—extending for 2 minutes 19 seconds along the line of the components but for only 30 seconds of arc across each component. Hence any answers about the angular size of this source assessed from a measurement along a single baseline might vary by 4 to 1 depending on the geographic direction of the baseline. The obvious way out of this difficulty would be to shift the remote aerial around and repeat the measurements along different baseline directions until a representative sample could be obtained from N–S to E–W directions. Jennison did this in his measurements of the Cygnus source but the difficulties in repeating this idea for a large number of sources with baselines extending across England need no emphasis.

To overcome this problem Rowson developed a system known as a tracking radio interferometer. In the initial tests the small broadside array of area 36 square metres which had been used in some of the earlier work was mounted on a steerable platform. It was first used at the airfield in Lincolnshire over the baseline of 61 100 λ*. With the Mk I telescope following a source across the sky this remote aerial was steered by hand to keep the beam on the source and the signals were transmitted to Jodrell Bank via the Holme Moss repeater station as in the work with the fixed cylindrical aerials.

Of course, with this system the baseline varies with the position of the source in the sky. For example, in the simple case shown in Fig. 7a if

---

* Tests of the system had been attempted at Holywell without much success because of the R.T. interference from ships using the Mersey.

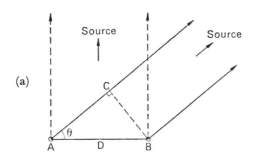

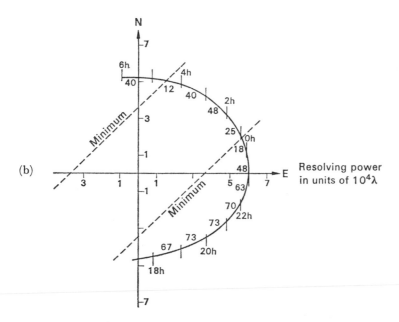

7. Effective baseline and resolving power, at different hour angles of observation, of a long baseline interferometer. The elliptical line in (b) is for a baseline D of 61 100λ and is the path followed by the tip of a vector from the origin, the direction of which gives the position angle of the baseline relative to the source 3C 295. The length of this vector is proportional to the effective resolving power in that direction. The figures outside the elliptical line give the hour angle of observation, while the figures within the line are computed values of fringe amplitude for the double source model proposed in Fig. 8(b).

A and B are the two aerials separated on an E–W baseline by a distance D (61 100λ in the Jodrell–Lincolnshire experiment), then this maximum baseline would be realized only when the source is in transit. This was the condition of the measurements already described with the aerials fixed. If now we consider the hypothetical case of a source moving across the sky in a plane defined by the zenith and the E–W line, then as the aerials follow the source the effective baseline at any angle θ to the horizon will be BC or $D \sin \theta$. In any practical case the geometry is more complicated because the sources studied will lie at various declinations and the constantly changing baseline BC will refer to varying position angles across the source.

In fact, it can be shown that with an E–W baseline the results are equivalent to observations of the source in the zenith with one aerial fixed and the other moving continuously along an elliptical path. The first measurements with this tracking interferometer were made on the source 3C 295 in Boötes and the ellipse showing the variation of resolving power at different hour angles of observation for this source is shown in Fig. 7(b). In the experiment two new technical problems arise. Firstly, the fringe speed varies as the source moves across the sky, and it was necessary to devise a system to vary the speed of the velodyne at a rate proportional to the declination of the source. Secondly, the compensating delay in the signal from the Jodrell aerial to allow for the transmission times from the remote site could no longer be fixed. A continuously variable mercury delay line was used to effect these adjustments with an additional quartz delay line for baselines greater than 25 km. The geometry and technical details of the equipment have been described by Rowson* together with some of the early results.

The initial experiment on the distant radio galaxy 3C 295 gave the results shown in Fig. 8(a) where the apparent flux is plotted as a function of hour angle.† It was found to be impossible to account for this variation of flux with hour angle on the basis of any single emitting centre. A source model with two components similar to the Cygnus source was found to give the best fit to the observations, and it was concluded that 3C 295 consists of two components separated by 4 seconds of arc in position angle 135° each component being less than 1 second of arc in extent along this line but 1·7 seconds of arc perpendicular to the line as shown in Fig. 8(b). The curve in Fig. 8(a) is the variation to be expected from this model with the flux from one of the components being two-

* B. Rowson, *Mon. Not. R. astr. Soc.*, **125**, 177, 1963.

† B. Anderson, H. P. Palmer, and B. Rowson, *Nature*, **195**, 165, 1962. The second minimum could only be observed after Messrs. Marconi of Chelmsford had kindly loaned us a second quartz delay line.

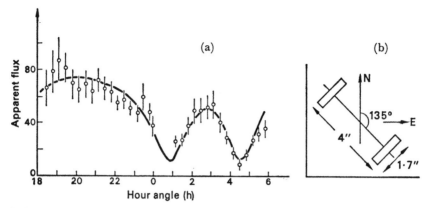

8. Observations of the radio source 3C 295 at different hour angles, with an interferometer of aerial separation 61 100λ. Each point is the mean value over 20 min with its standard deviation, (b) shows the two-component source model giving the best fit to (a).

thirds of that from the others. Hence, this distant radio galaxy was found to bear a close structural resemblance to the Cygnus source as far as the radio emission was concerned. The calculated linear dimensions of each of the components on the basis of the measured redshift was about 8 kpc, or about one quarter of the projected size of the components in Cygnus.

The resolving powers of better than a second of arc achieved in the Jodrell Bank–Lincoln observations had been insufficient to resolve 5 of the radio sources of the 384 studied. The fringe amplitudes of these sources had remained the same at all the baselines and it could be said only that their angular sizes must be less than 0·8 seconds of arc. The work had received stimulation from the unexpected discoveries of the quasars and the possibility of distinguishing between cosmological models from the analysis of the diameters remained as an important issue. In that period of 1961–2, when considering the next stage there were several baffling technical problems. Firstly, the work with the tracking interferometer had confirmed the deduction from the analysis of the 384 sources that the angular structure showed considerable asymmetry in a large proportion of the sources. Hence transit observations with fixed aerials must be superseded by a tracking system over a range of hour angles so that one could be more certain what the measurements implied in relation to the overall structure of the source. Secondly, a further increase in resolving power was desirable, but we were already in Lincolnshire and a further doubling of the baseline towards the east would take us into the North Sea. Thirdly, the 384 sources already studied were the most intense known. Only 5 remained unresolved and hence an increase in sensitivity was necessary in order to bring more sources within the scope of the measurements.

The desirability of tracking meant that we would have to use aerials smaller than the cylindrical paraboloids at the remote site and initially the various needs seemed mutually contradictory and an immediate step forward proved impossible. Furthermore, we could only use the Holme Moss site during the summer and other arrangements would have to be found for a more extended and consistent programme. Fortunately at that time parametric amplifiers were just becoming available which gave worthwhile receiver sensitivities at much higher frequencies than that of the 158 MHz so far used. We had a reserved band at 408 MHz and Anderson* pointed out that, using parametric amplifiers, sufficient sensitivities could be obtained with a remote aerial small enough to be steerable. This would also solve the problem of the baseline. Even with the existing separation to Lincolnshire the effective baseline would be increased by $2\frac{1}{2}$ times by using the higher frequency.

In the face of these various problems we decided, after the completion of the Lincolnshire measurements in the summer of 1961, to develop a system on 408 MHz using a remote steerable aerial which could be controlled by a radio link from Jodrell Bank. In order to get sufficient sensitivity the remote aerial would have to be larger than the array of dipoles used on the tracking aerial in Lincolnshire. We decided that it would be possible to build a 25 ft diameter paraboloid in our own work-shops out of aluminium tubing and one of the new students, Donaldson,† was given this task, the criteria being that it had to be accurate enough to work on 408 MHz and it was to be mounted on the rotating structure of an old defence radar. Meanwhile Rowson was to develop the remote radio control using the existing 158 MHz steerable array.

In due course another disused airfield was found close at hand— fifteen miles to the south-west of Jodrell at Wardle near Nantwich.‡ These various plans proceeded while the analysis of the Lincolnshire results were being completed. The radio link to Wardle was successfully developed, some test results were obtained between the Mk I and the 158 MHz array at Wardle automatically controlled from Jodrell Bank.

---

* Dr. Brian Anderson came to Jodrell Bank as a research student in 1960 after graduating in the University of Manchester. In 1967 he was appointed to a Lecture-ship in Radio Astronomy.

† Dr. Wilfred Donaldson, a graduate of Liverpool University, was a research student at Jodrell Bank from 1962 to 1969. After a period in the University of Lan-caster he joined the Department of Inland Revenue.

‡ Discovered by Rowson on a cycle ride, this airfield was to figure prominently in our subsequent history—see Ch. 7.

Anderson developed the 408 MHz receiver system using the parametric amplifier and Donaldson toiled in the workshop on the paraboloid. During the summer of 1963 these various developments approached completion and the intention was to move the paraboloid to Wardle and settle down to a long study of the structure of the radio sources already measured in transit between Jodrell and Lincolnshire.

The programme never materialized. In the summer of 1963 I went to the Soviet Union as the guest of the Academy of Sciences. After some days in Moscow during which the President of the Academy, Academician M. V. Keldysh, informed me that I was to visit their deep space tracking station, I was taken on 27 June to the Crimean Astrophysical Observatory. There was a summer school for young Soviet astronomers in progress. The next morning I lectured to the astronomers and the various proceedings continued all day with the drama at the end. As the heat of the afternoon was subsiding I. S. Shklovsky gave a two-hour display of his virtuosity on the subject of the quasars. He discoursed in detail on the interpretation of the few quasars identified. For 3C 286 he maintained that the absence of recognizable lines or bands in the emission spectrum was because they were redshifted out of sight and he estimated the distance of the source as 4000 megaparsecs. The next morning Shklovsky arranged a private and prolonged discussion with me during which he spoke with such excitement that three interpreters were necessary (although normally his English is very good). He was convinced that the energy sources in the quasars were of exceedingly small dimensions (and he was eventually proved to be right). Our most recent results, which I had discussed with him, showed only that 5 out of the 384 sources had angular dimensions less than 0·8 seconds of arc. Shklovsky pleaded with me to find some means of improving still further the resolving power since he was convinced that in cases like 3C 48 we would find that the energy producing region of the source was not more than 0·15 seconds of arc. To the purely cosmological argument for improved resolving power. Shklovsky, with characteristic insight, had added the physical argument.

There were at least two results of these encounters. The first was that I was impressed with the urgent need to extend the baselines of our interferometers still further. The second was that we were to explore the possibility of using a very long baseline between Jodrell Bank and the deep space tracking telescopes in the Crimea which, I was surprised to find later, Shklovsky was able to use.

After this encounter it was hardly surprising that when I returned to Jodrell in mid-July I told Palmer to cancel the arrangements for taking the small steerable telescope to Wardle and produced the Shklovsky arguments for a further increase in resolving power. Even at the Lincoln site the new frequency, 408 MHz, would give us an improvement of 2·7 times in resolving power. However, the Lincoln site was unusable in the winter and we wanted an even longer baseline. We studied the map of England and Palmer and Rowson set out for Yorkshire. If they could find a site there, the G.P.O. mast at Windy Hill could serve as the repeater station. Furthermore, unlike the B.B.C. mast at Holme Moss, we would be able to use it all through the year. With the help of the local Electricity Board, Palmer soon located a helpful farmer at Pocklington, 132 km to the north-east of Jodrell Bank, which at the frequency of 408 MHz (0·73 m wavelength) would give us a baseline of 180 000 wavelengths and three times the resolving power compared with the previous measurements from Lincolnshire.

The transport and setting up of this remotely controlled device during the winter months was an arduous process but before the end of the summer of 1964 Palmer and his group had used this system to study 34 radio sources, which comprised all those of sufficient strength which had either given fringes at the longest baselines previously used, or had been provisionally identified as quasars.* Sixteen of these sources still gave measurable fringe patterns and 5 showed no sign of resolution implying an angular diameter less than 0·3 seconds of arc. Of these 5 sources 2 had already been identified as quasi-stellar sources—including 3C 286, the object about which Shklovsky had been so excited and for which he had assigned a redshift of $z = 0.86$.† The other 3 unresolved sources, CTA 21, CTA 102, and 3C 119, were at that time unidentified.‡

The impact of the 1964 Solvay Conference

I have already referred to the stimulating effect of the 1958 Solvay Conference in Brussels, Six years later it was my good fortune to be invited to the 13th Solvay Conference held in Brussels during September of 1964 on 'The Structure and Evolution of Galaxies'. Much had already been said at the Conference on the energy problem by Hoyle and Fowler, by the Burbidges, by Schmidt and Sandage. When we began the last session, the Chairman, J. R. Oppenheimer, said:

In this session we will discuss topics about which we know very little. Perhaps then we will all be treading on common ground. We are going to discuss

* B. Anderson, W. Donaldson, H. P. Palmer, B. Rowson, *Nature*, **205**, 375, 1965.

† A redshift of 0·849 was subsequently confirmed for this quasi-stellar object. The other, 3C 287, transpired to have an even greater redshift of $z = 1.055$.

‡ In December of 1964 Sandage announced the identification of 22 more quasars amongst which was CTA 102 with a redshift of $z = 1.037$. The other two, CTA 21 and 3C 119, remain unidentified.

possible sources of very large energies—which one detects, and the still larger energies which one infers from some of the greatest radio-emitters . . . The obvious candidates for great energy are nuclear energy and gravitational energy; neither has proved to be the answer to the question of origin of these great sources, nor have we got proof that these are not adequate. But simply, they do not work because of the energy amounts, the time scale of their release, and the forms of energy they release. Gravitation has never been tested under these conditions, and I keep an open mind about the fact that there could be great surprise here which would not in any way contradict the views that Einstein had about gravitation, except that they might not be as simple as he stated. I thought this afternoon we would have some speculation on possible sources of energy. Professor Alfven has worked very ingeniously on the production and annihilation of matter and anti-matter which is obviously a very tempting source. I think Ambartsumian has the suggestion of two colleagues Zeldovitch and Novikov, to discuss briefly.*

I had been able to report the new results of Palmer which underlined the difficulty, that it was necessary to account not only for vast energies but also for the production of that energy in small volumes of space. Apart from the overall cosmological problem which remained unsolved there was now this additional problem of the physical processes which could produce the energies observed. Suggestions were made that the energy-producing regions in the quasars might be far smaller than the limits already set in the existing angular diameter measurements.

I returned to Jodrell with a new determination to persuade the interferometer people to increase still further their baseline. It was natural that Palmer and his collaborators who had, with such effort, established the Jodrell–Pocklington link, wanted to build up more results at that baseline. I wanted to pursue the four sources which still showed no sign of resolution at that baseline. Everybody thought of the difficulties. The 25 ft dish used at Pocklington was not big enough. Then why not try to borrow the large telescope at Fraserburgh in Scotland which the Americans had built many years ago for some defence research? I was told that there would be too many radio links and that we could scarcely keep a single repeater in action. Then why not reduce the wavelength if we could not increase the baseline?—the Jodrell telescope would work adequately on a wavelength of 21 cm. This was impossible because the 25 ft dish was no good on 21 cm. But R.R.E. at Malvern had very good radio telescopes which would work down to 11 cm easily. That suggestion left the critics temporarily silenced during which time I drove to

* *The Structure and Evolution of Galaxies*, Proceedings of the 13th Conference on Physics at the University of Brussels, September 1964 (Interscience Publishers 1965), p. 168.

Malvern and talked to J. S. Hey and George MacFarlane* for an hour about the tremendous scientific issues which might be involved in these measurements. My own conviction about the importance of achieving this greater resolving power is indicated by the fact that although I had not returned from Brussels until the Saturday evening of 26 September, I had gone through these internal arguments and was in Malvern on Friday, 2 October. I had no doubt at all that they would react favourably and in that I was not disappointed. In that manner began a long and fruitful collaboration.

This was a moment of rare opportunity when the hard-pressed interferometer team was temporarily aided by a visiting Professor—A. D. Frost—from the University of New Hampshire and the R.R.E. team by an experienced radio astronomer on leave from the Australian C.S.I.R.O.—Dr. O. B. Slee. The technical difficulties of the experiment rapidly evaporated before the onslaught of these two groups and by May 1965 measurements were in progress between the Mk I telescope at Jodrell Bank and one of the 82 ft radio telescopes at Defford, operated by the R.R.E.

Although the physical separation of the two telescopes was slightly less than in the Pocklington experiment (127 km as against 132 km) the wavelength was 21 cm instead of 73 cm; here the resolving power was three times as great. (The maximum effective baseline was 600 000 λ). There was no conveniently situated mast on a high hill between Jodrell Bank and Malvern and two repeater stations had to be established for the microwave link system. The fringe frequency varied with hour angle from zero to 40 cycles per minute. In order to display these high fringe speeds on a chart recorder an almost identical frequency was produced continuously by a digital device, and subtracted from the interferometer output so that the fringe pattern on the chart was normally slower than 0·5 cycles per minute.†

All the 5 sources previously unresolved gave fringe patterns. The pattern for CTA 102 was constant at all hour angles and hence its diameter was less than 0·1 second of arc in all position angles. The fringe amplitude from the other four quasars (3C 119, 3C 286, 3C 287 and CTA 21) varied with hour angle indicating that each source was less than 0·1 second of arc in at least one direction and that their maximum dimensions were in the range 0·12–0·16 second of arc. These results were

* Dr. G. G. MacFarlane, C.B. (now Sir George MacFarlane) was then the director of the Royal Radar Establishment, Malvern. In 1967 he was appointed Controller (Research) Ministry of Technology.
   † See note, p. 51.

published* in October 1965—almost exactly a year after my visit to Malvern. Assuming the redshifts to be cosmological the results showed that the linear dimensions of the radio emitting regions must be only a few hundred parsecs in the case of CTA 102, 3C 287, and 3C 286, and that the components of 3C 273 were smaller than 170 parsecs. Already it was clear from these results that we must be concerned with processes capable of emitting an energy equivalent to millions of suns but contained within a spatial volume minute in comparison with galactic dimensions.

I cannot recall that at this stage any of the participants argued against the next stage—to operate with R.R.E. using the Mk II telescope which could lead to a further trebling of the resolving power over the same baseline by using a wavelength of 6 cm instead of 21 cm. Of course, there were problems. The Mk II was much smaller than the Mk I, but with the parametric amplifiers then becoming available it seemed possible that a useful number of results could be obtained. The fringe speed machine needed a further significant improvement to reduce the fringe speed on the recorder to a suitable rate.† These improvements were made while further measurements were in progress between the Mk I and Malvern on 21 cm. Eventually, during 1966, the goal of a million wavelength baseline was achieved by operating on 11 cm with the Mk II and Malvern and finally of two million wavelengths by operating on 6 cm. Altogether 50 sources were studied. There were 46 which were unresolved in the 21 cm experiment, 25 were quasars, 7 radio galaxies, and the others unidentified. 18 of the quasars, 4 of the galaxies, and 9 unidentified objects were found to be less than 0·05 seconds of arc. In the limited observations which were possible at 6 cm wavelength 3 quasars and one Seyfert Galaxy were found to have dimensions of less than 0·025 seconds of arc.

The authors concluded their published account‡ with these words:

The interferometer observations described here demonstrate the remarkably high angular resolution that can be achieved at radio wavelengths. The results place new upper limits, by direct measurement, on the angular dimensions of the important class of radio sources characterised by small size

* (From R.R.E.) R. L. Adgie, H. Gent, O. B. Slee; (from Jodrell Bank) A. D. Frost, H. P. Palmer and B. Rowson: *Nature*, **208**, 275, 1965.

† In the first of these improvements the Meccano system described in Ch. 4 was replaced by a tape-recorder producing signals according to a train of digits prepared on an Argus 100 computer. In the second improvement the recorder was replaced by a solid state device which solved the appropriate equations.

‡ (From Jodrell Bank) H. P. Palmer, B. Rowson, B. Anderson, W. Donaldson, G. K. Miley; (from R.R.E.) H. Gent, R. L. Adgie, O. B. Slee, J. H. Crowther; *Nature*, **213**, 789, 1967.

and exceptionally high brightness temperature. Perhaps the most surprising result of the observations is the discovery that such a large number of sources have these small dimensions, and that this may be a general feature of most of the quasars. Any theory to account for these sources must explain the large radiating power being generated from such a small volume. For example, assuming the usual cosmological interpretation of their redshifts, the upper limits to the linear dimensions of the quasars 3C 273B and the Seyfert Galaxy 3C 84 are about 40 and 6 parsecs, respectively.

The distance of the nearest star to the Sun is 1·3 parsecs. The previous measurements had indicated that the spatial volume in these remote objects in which such vast amounts of energy were generated were minute compared with galactic dimensions. These new measurements showed that the spatial volumes involved were to be compared only with volumes of space containing a few of the nearer stars in the solar neighbourhood.

For more than a decade we had led the international field in the search for higher and higher resolving powers. With these remarkable results we reached our climax. Others had been stimulated by this work to seek various technical solutions to the problem of multi-million wavelength baselines—and in this work we had neither the manpower nor financial resources to compete in their time scales.

6

Independent tape-recording interferometers

In Chapter 5 I have described how my conversation with Shklovsky in the Crimea in July 1963 stimulated me to urge Palmer and his group forward to greater resolving powers. Even before I left Russia I had discussed with Shklovsky the possibility of using the radio telescopes of the deep space tracking station in the Crimea linked with Jodrell Bank for a great extension of the existing baselines. I promised that a memorandum would be prepared for their consideration. This, with the authorship of Anderson, Palmer, and Rowson was despatched to the Soviet Union in November of that year. It proposed a linkage of the Crimean and Jodrell telescopes on a wavelength of 42 cm. With a separation of 2 200 km this would be a baseline of 5 million wavelengths—a hundred times greater than any interferometer used up to that time.

The major problem was, of course, to link the two telescopes. At that time only one repeater station had been employed in the Jodrell work. The memorandum pointed out that fifty such links would be necessary between the Crimea and Jodrell Bank. This possibility was dismissed on two counts. First the cost—the hire of only one line of commercial telephone quality between the U.K. and the U.S.S.R. cost one rouble (equal at the time to seven shillings) per minute. Apart from the cost, it was doubtful if such a link would be satisfactory because the delay in each section would have to be known to a ten-thousandth of a second.

A new proposal was therefore made. This was to make tape-recordings separately at the site of each telescope and subsequently bring the tapes together and correlate the signals. The memorandum discussed the two possible types of interferometer—the post-detector correlation version developed by Hanbury Brown (see Chapter 2) which had been abandoned by us for radio measurements because of its insensitivity, and the phase coherent version, which we had subsequently used. The former was much easier to consider for this experiment, placing restrictions on bandwidth, stability, and synchronization which were not too severe.

However, the memorandum did not recommend this because with the available sensitivity it seemed that only one or two of the hitherto un-resolved sources could be studied. The recommendation was to develop a phase correlation system using high quality tape recorders, wide bandwidths, and precise timing and synchronization systems.

The timing restrictions were severe. The independent local oscillators in the Crimea and Jodrell Bank would have to be stable to one part in 10^{10} per day and the alignment of the recordings would have to be made to better than a millionth of a second. A variety of proposals were made for the achievement of this synchronization, such as the use of radar echoes reflected via the Moon, or the use of pulsed transmissions from an Earth satellite, or, if clocks of sufficient accuracy and stability were avail-able, the transportation of one clock between Jodrell Bank and the Crimea by air.

The subsequent history of this proposal for co-operation is curious in the extreme, particularly when judged against the background of the other arrangements made during my 1963 visit. Essentially, in the astro-nomical domain, I made three collaborative arrangements with the Soviet Academy. These were the interferometer experiment mentioned above, the flare star observations, and a bistatic radar experiment on the planet Venus. The third of these described in Chapter 14, was by far the most difficult, organizationally and technically. The experiment in-volved the Soviet deep space facility transmitting to the planet Venus, with reception at Jodrell Bank after the radar pulses had been reflected from the planet. When it is remembered that the Soviets had a tight security barrier around their Crimean equipment and that its primary purpose was the control of deep space probes it now seems to me re-markable that this experiment was ever completed. On the second pro-posal, no problems ever arose and the collaboration continues today (see Chapter 13).

The arrangements with the Soviet Academy were that I should send the proposals to Academician Kotelnikov, the Director of the Institute of Radio Technics and Electronics in Moscow, the institution responsible for the Crimean installations.* On 15 November 1963 I wrote:

Dear Academician Kotelnikov, You will recall that during my visit to the Soviet Union we had discussions about the co-operative programmes be-tween Jodrell Bank and your establishment in the Crimea. On the 19 Septem-ber I sent you the detailed proposals for the bistatic radar experiments on the planets, and I am now sending you the other memorandum on the feasibility of using our two telescopes as an interferometer over a very long baseline. It

---

* Also, as agreed, I sent copies to Professor Alla Massevitch, the Secretary of the Astronomical Council of the U.S.S.R.

seems to me that both documents, particularly the one sent to you previously on the radar experiment, need a discussion between the experts on the two sides in order to initiate the programmes. As I said in my previous letter, we would be happy to arrange for this to take place either at Jodrell Bank or we would send representatives to Moscow. I hope very much that some progress can be made.

The memorandum by Palmer and his collaborators was a comprehensive one, containing the results of much hard thinking and discussion and—as far as we were aware—it made the first proposal to use tape-recorders at the two ends of the interferometer and bring them together for subsequent correlation. It was our assessment that, with the specialized tape-recorders of very high quality and stability which had by that time been developed for recording of TV transmissions, and by using the rubidium atomic clocks which were becoming available, the experiment was entirely feasible and might well quickly lead to measurements of great importance.

Immediately after my return from Russia in July 1963 I had started negotiations for Palmer to visit the Soviet Union. Dr. Moiseev, a young radio astronomer from the Crimean Astrophysical Observatory had worked at Jodrell Bank in Palmer's group and he was now the head of the radio astronomy section at the Observatory. During my visit he took me to the beach at Katziveli where a steerable radio telescope was being constructed. It was natural that Moiseev should be anxious for Palmer to make a return visit with the possibility of assisting in the initiation of the radio telescope. Professor Severny, the director of the Observatory, also pressed me to arrange this visit. The negotiations began straight away on the basis of the arrangements for mutual visits between the Royal Society and the U.S.S.R. Academy of Sciences.

Although I had initiated this in July through the Royal Society the end of the year approached without further news either of Palmer's visit or of the fate of his memorandum. I asked Dr. Garrett, our Scientific Attaché at the Embassy in Moscow if he could find out what was happening.

Garrett to Lovell 12 December 1963: I took the opportunity when visiting the Foreign Department of the Academy of Sciences yesterday to enquire what progress had been made in the proposals (a) that Dr. H. P. Palmer shall spend about one month in the Spring of 1964 at the Crimean Astrophysical Observatory to work with Dr. Moiseev and (b) for joint observations between Jodrell Bank and the Crimean Astrophysical Observatory. With regard to (a) there appears to be no objection in principle, especially since foreign astronomers have previously worked at the Crimean Astrophysical Observatory. Dr. Korneev, the Head of the Foreign Department, promised to look into the matter and maybe this will result in some further progress being made. Little information was available regarding (b). As you predicted the question has been referred to the Academy's Astronomical Council and as far as I could gather the matter was still under discussion.

On the basis of this letter I asked the Royal Society to send a telegram to Korneev suggesting that Palmer should arrive in Moscow on 27 May for eight weeks. Lovell to Garrett 18 December 1963: 'The main point of Palmer's trip will be to visit Moiseev at the Crimean Astrophysical Observatory, but I am most anxious that he should also discuss the memorandum which we have already submitted to the Academy, dealing with the proposed joint observations between the Crimea and Jodrell Bank for the measurements of the angular diameter of radio sources.'

Garrett to Lovell 10 January 1964: The Ambassador called yesterday on the President of the Academy of Sciences and took the opportunity to ask Keldysh whether any progress had been made in the consideration of your proposals for joint observations between the Crimea and Jodrell Bank and for Palmer to meet Moiseev at the Crimean Astrophysical Observatory. Keldysh replied that the proposals were being studied in detail by the experts on their side and mentioned in particular Academician V. A. Kotelnikov and the Institute of Radio Electronics. Korneev the head of the Foreign Department, affirmed that the Academy had no objection in principle to the proposed visit of Palmer. There has not therefore been any great positive advance.

However, by the end of the month the Academy at last agreed that Palmer's visit should take place in the spring, and I continued to press the point that he should have discussions on the memorandum in addition to the liaison with Moiseev. In April, quite unexpectedly, I had a pleasant letter from Professor Gabriel Khronov (dated 4 April 1964) who sent me a copy of a scientific paper 'in memory of our meeting in the Crimea last summer'. The paper dealt with measurements of the radio emission from planetary nebulae made in the Crimea. Professor Khronov's letter ended 'It may be of interest for you to know the situation with your propositions on the long-base interferometer. I can inform you, unofficially, indeed, that Project is under investigation now, and as much as I can guess, the situation is rather favourable. But, let's wait. Here you ought to show patience, because the time constant may be a significant one.'

Professor Khronov was, like Shklovsky, a member of the Sternberg Astronomical Institute in Moscow and it was evident from this letter that our memorandum had reached Shklovsky's group. At least in my reply to Khronov (15 May) I was able to say that Palmer was arriving in Moscow on 28 May: 'I expect he will spend most of his time at the

Crimean Astrophysical Observatory but he is certainly hoping for an opportunity to have further discussion on this co-operative programme.'

That, in fact, is exactly what happened. Palmer spent most of this visit in the Crimea, either at the main Observatory or with Moiseev at the site of the 22 m paraboloid near Simeiz (which was still not working). He visited the Sternberg Institute in Moscow on two occasions.

Palmer to Lovell 27 May 1964 (written from the Hotel Ukraina, Moscow): I was met by Mr. Garrett, a lady from the Astronomical Council and by Intourist with 2 too many cars and 2 reservations! This afternoon I visited the Sternberg Institute and met Prof. Shklovsky, Dr. Slish, Karadashov and Scholomusky. . . . [The letter then gave details of further work in the U.S.S.R. which led Shklovsky and his colleagues to maintain that they had studied some objects which they believed to have extremely small diameters perhaps only 0·005 seconds of arc.* On the subject of the proposed long baseline interferometer the letter continued] They hope to operate between Simeiz and the Space Aerial at $\lambda = 10$ cm. The distance is about 50 km, so that the separation is approximately 500 000 wavelengths. The suggestion arose during your visit! With the prospect of even more very small sources, they are really very keen on a co-operative experiment. They said they agreed with our report, of which they had an English copy. They were in favour of wide bandwidths. They definitely favoured tape recording and were not very concerned about timing up the tapes during the observations. They felt that the computer should search for the correlation, but it was to be our computer, I think. They felt that the higher the resolution the better, an increase in fringe speed should help to avoid trouble with tape flutter. They stated that the high resolution should be approached progressively, with which I agree, and suggested that we should separately attempt to develop a tape recording system for test on shorter baselines. At the Space Station there is negligible chance of changing frequency. They say astronomy gets very little time anyway.

Later Palmer wrote to me from Simeiz where he was staying with Mosieev and trying to help him reduce some observations which Moiseev had made previously during a visit to Jodrell Bank. 6 June 1964: '. . . also he [Moiseev] arranged 3 meals each day in an excellent resthouse, 10 miles away, on rough, slow coastal roads. So when 60 miles = 3 hours motoring, plus 3 meals plus bathes was subtracted from each day, little time remained for analysis. Now he has agreed to reduce this to two even larger meals per day.' One of the reasons for arranging Palmer's visit at this time was in the expectation that he might be able to help Moiseev begin observations with the radio telescope. However

. . . his telescope is not yet complete, and probably looks much as you saw it. I gather the bowl structure has been externally adjusted for shape, but is not yet surfaced. They are trying now to lift the base, 50 tons, to insert a wooden

* They were eventually proved to be quite right in their prediction.

cable twister but yesterday's efforts were not successful. The bowl will weigh 50 tons, 22 metres (diameter). The Control room is still empty except for potted plants. He talks about completion in 1964, but without too much conviction, as he also talks of a visit to Australia for 6 months. . . . He says there is a penalty clause in this telescope contract, but as he is not short of money, and the Leningrad Contractors also have plenty of roubles, it is not very effective.

In this letter Palmer wrote that in his opinion it would be better to carry out the interferometer experiment with Moiseev's telescope when it was complete rather than with the much larger deep space facility which I had proposed because 'it would allow us to change the wavelength readily, at the cost of a small reduction in collecting area'.*

I have described in the previous chapter that shortly after Palmer returned from the U.S.S.R. I attended the Solvay conference and returned to initiate the long baseline work with R.R.E., Malvern. All our available efforts were therefore devoted to achieving the measurements over a baseline of a million wavelengths. It was simply not possible to raise further effort or enthusiasm for the development of the tape-recording interferometer at that moment. Hence, as far as we were concerned the project did not make much further progress until the President of the Soviet Academy of Sciences, Academician M. V. Keldysh, with the official party of distinguished Soviet Scientists, visited Jodrell Bank on 17 February 1965.† I was handed a letter from Shklovsky:

In accordance with the existing agreement, there is a work proceeding at the Sternberg State Astronomical Institute on the elaboration of a transcontinental radio interferometer. We have studied the papers you have sent, and we do agree with conclusions drawn by you and your collaborators, but the single exception concerning the bandwidth. On our opinion, an experiment of such a scale is worth undertaking, only if sensitivity is high enough. So, we think that interferometers of this type should have a bandwidth of recorded signal at least as broad as 100–200 kc/s. We believe, that the difficulties then arising would be possible to overcome. At present, we are studying the possibility of construction of a broad band magnetic recorder with a proper stabilization of the speed of pulling. We would be glad to know your

* Palmer was probably correct in this judgement. It was with this telescope that the Americans collaborated in 1970 and our own ineffective progress in the discussions was probably exacerbated by the secrecy surrounding the telescopes of the deep space installations.

† The visit, which was under the official auspices of the Royal Society consisted of Academicians M. V. Keldysh, N. N. Bogolyubov, N. N. Semenov, M. M. Sisakayan; Professor N. F. Krasnov, Mr. N. A. Plate, Dr. V. A. Filippov, Mr. S. A. Sokolov, and Mr. V. S. Vereschetin. The President, Keldysh, and the Biological Secretary of the Academy, Sisakayan, stayed privately in our house—a rare honour and act of trust at that time.

opinion on the possibility of our co-operation in the construction of such a recorder. With best personal regards, sincerely yours, I. Shklovsky.

With this evidence of continued interest on the Soviet side it was clearly high time we rationalized our own ideas and made some kind of progress in spite of the other preoccupations.

The technical problem

There was no easy solution to the technical problems posed by our suggestion in the memorandum to use a tape-recording system. We agreed with Shklovsky about the bandwidth—this was already accepted in our memorandum. We were aware also, of the problem posed by his phrase 'proper stabilization of the speed of pulling'. These issues were wrapped up in the following considerations. Today, the subsequent development of video tape-recorders for commercial television and the availability of atomic clocks make them readily soluble. In 1965 the problems were formidable.

1. If C is the tape speed past the recording head, and G the wavelength recorded on the tape then the width of the gap between the poles of the magnetic recorder determines the usable bandwidth Δf. Since $\Delta f . G = C$, G must be greater than the gapwidth. For reasonable sensitivity Δf should be more than Shklovsky suggested—say 1 MHz, then as an example, if the tape speed is 100 in s^{-1}, G is 10^{-4} in and the gap must be less than one ten-thousandth of an inch.

2. With a 1 MHz bandwidth the path differences must be maintained in the play-back for correlation to the order of a microsecond. The best recorders available in 1965 had nothing like this speed constancy of the tape past the head. Indeed the jitter introduced by the uneven speed was of the order of 250 microseconds. It was this problem to which Shklovsky specifically referred in his letter.

3. A high degree of frequency stability is required for the fringe speed control. The frequency standard which is used to control the local oscillators must be such that $\Delta\omega . \Delta\tau < 1$ where $\Delta\tau$ is the integration time, and $\Delta\omega$ is the uncertainty in fringe frequency (that is, the relative uncertainty of the local oscillator frequencies). The integration time $\Delta\tau$ should be not less than 10 seconds otherwise the minimum signal to noise requirements will not be achieved. For a frequency f the above relation can be written

$$\left|\frac{\Delta f}{f}\right| 2\pi f \Delta\tau < 1$$

where $\left|\dfrac{\Delta f}{f}\right|$ is the stability of the frequency standard. The maximum
usuable frequency f, for $\Delta \tau = 10$ s, is then given by

$$f = \frac{1}{2\pi\Delta\tau}\left|\frac{f}{\Delta f}\right|$$

For a crystal oscillator $\left|\dfrac{\Delta f}{f}\right|$ can be 1 in 5×10^{12} for a few seconds,
but the ageing over a day reduces this to 1 in 10^{10}. Since observations
must be made over many hours on a single source this implies a
maximum usable frequency

$$f = \frac{1}{2\pi\Delta\tau}10^{10} = 160 \text{ MHz}$$

This is not high enough. A frequency of over 700 MHz was envisaged for the Soviet–Jodrell measurements and hence an improvement of ten times is needed. The only reasonable way of achieving
this was to use a rubidium frequency standard which ages by only 1
part in 10^{11} per month.

4. Apart from this problem of frequency stability there is also the
question of the actual timing accuracies on the separate tapes. When
played back for correlation it was considered that uncertainties of not
more than 1 microsecond over a day could be allowed. Within this
uncertainty bracket a search between two tapes for correlation was
considered to be feasible. Since there are $8\cdot4 \times 10^{10}$ microseconds in a
day, this implies a constancy of about 1 in 10^{11} per day—a criterion,
which again could be met by a rubidium clock but not by a crystal
clock.

Whereas a rubidium standard would meet the requirements in 3 and
4 there were the stringent requirements on tape-recorders set by 1 and 2.
In 1965 only the exceedingly expensive video tape-recorders (such as
used by the television companies) could be considered. The scheme envisaged was to use a servo system to detect the timing marks on the tape
from the rubidium standard and then apply the appropriate signal so
that the tape speed servo controls would pull in the tape speed appropriately.

By the early summer of 1965 Anderson and Palmer had studied these
issues sufficiently to define their minimum requirements. They said that
it was essential to purchase two recorders and two rubidium standards
and that the cost would be at least £50 000. Sums of that order were
simply not available. However, by the autumn of that year we managed

to find £5000 to buy a rubidium standard—which we urgently needed in any case for other reasons—and we decided to make an application to the S.R.C. for a tape-recorder and other equipment which at least would enable a start to be made on the system. As part of our Consolidated Grant application* for 1966 we asked for £12 500. After explaining the development and importance of the work already in progress between Jodrell Bank and Malvern the application went on:

These reasons suggest that baselines greater than 1000 km will probably be needed before the angular structure of the radio emitting regions associated with quasars is adequately understood. It might just be possible to extend our present techniques sufficiently to give reliable operation with baselines of this length, using many microwave links and correlation in real time. An alternative and more attractive scheme is to make tape-recordings simultaneously at each telescope, and search for the correlation subsequently, when the tapes are replayed at a computing centre. A completely independent and fully operational interferometer would require two rubidium or hydrogen maser frequency standards, and three wide band video tape-recorders (if it is assumed the computer is near one of the telescopes, so that one of the tapes can be recorded and subsequently reproduced on one instrument). However, the system can probably be tested and proved using only one 'portable' tape-recorder at the distant telescope, and a VHF phase locking signal instead of frequency standards and other existing equipment. Application is therefore made for authority to purchase one tape-recorder, sufficient tape for 24 hours of observation and the buffer core-store and associated electronics required to correct misalignment of the tape during reproduction. If, as seems likely, there is still a pressing need for longer baselines, it will be possible to introduce frequency standards and additional tape-recorders. It would then be technically possible to perform co-operative observations with most of the major radio telescopes on Earth.

This application made in September 1965 was for the calendar year beginning 1 January 1966. A year later (September 1966) my application for the 1967 grant contained the following.

Satisfactory progress has been made with the development of the long baseline interferometer technique. . . . The wide band tape-recorder which is required to test the interferometer system will be delivered in October 1966. The digital buffer store and other associated electronics are nearing completion and it is expected that preliminary tests will have been completed satisfactorily by the end of December. Both the recent co-operative observations, at

---

* The Science Research Council introduced the idea of the 'Consolidated Grant' for large research departments, in order to avoid the processing of a multiplicity of small applications and to give some reasonable freedom of manoeuvre to the applicant. At that time our grant was running at about £60 000 per annum, but more than £26 000 of this was for the continued support of research and technical staff. The amount available for new equipment was about £33 000 on which there were heavy demands for the new developments in digital data processing affecting nearly all our researches.

wavelengths of 21 and 11 cm between Jodrell Bank and R.R.E. Malvern, and also recent analysis of radio scintillation phenomena have shown that at least 12 quasars have radio-emitting regions smaller than 0·06 arc sec so that the need for higher resolving powers is more pressing than ever. Application is therefore made for authority to purchase a second wide band tape-recorder, a rubidium frequency standard for use at the remote site, further supplies of magnetic tape and some additional electronic equipment.* These instruments would enable the full system to be tested and used as an operational inter-ferometer which is completely independent of radio links. Observations could then be made with greater telescope separations than have been possible so far, permitting co-operative programmes with more distant observatories having radio telescopes in the British Isles or, indeed, anywhere on Earth.

The application was dated 29 September 1966. The £18 700 for this equipment was well over one-half of the money £32 750) for which we could apply for apparatus in that grant. With the £12 500 already spent we would still be £20 000 short of the estimate made by Palmer and Anderson in 1965. During the initial discussions which led up to the framing of this new application it was evident to us all that we could neither meet the total cost of this tape-recording equipment nor could we bring sufficient staff to the problem of mounting such a complex operation.

Our work with R.R.E. was going extremely well and one day before the application was despatched I had a meeting with Dr. J. S. Hey and Mr. H. Gent of R.R.E. Palmer's minutes of the meeting began by stating that

Prof. Lovell recalled that we had now completed two seasons of co-opera-tive interferometric observations between Jodrell Bank and Malvern and he regarded the measurements made during this programme as being amongst the most interesting and useful with which Jodrell Bank had been associated. The whole series seemed to him to be a most fruitful example of well planned co-operative work. As a result of these measurements . . . it seemed clear that great increases in resolving power would be needed in order to measure the structures of these quasars. He had therefore asked for this meeting in order to invite R.R.E. Malvern to join with Jodrell Bank in future programmes to obtain still higher resolving powers.

My first point was to propose measurements between the two sites on the OH emission at 1660 MHz, since it was already known that the sources of the emission were smaller than 1 minute of arc. Hey agreed and did not foresee any difficulties.

* £11 500 for the recorder (an Ampex FR 1800 H); £3700 for the rubidium standard, £2500 for 40 reels of magnetic tape and £1000 for associated electronics.

Professor Lovell then explained that his main reason for calling this meeting, however, was to discuss the development of interferometers with baselines greater than 1000 km; this aim might be achieved using interferometers in which the radio links were replaced by independent local oscillators derived from separate frequency standards, and wide band tape-recorders. He explained that two other groups in North America, at Penticton* and at Green Bank†, were planning similar experiments. He did not regard this as any sort of race‡, however, but mentioned these other programmes as reinforcing his own conclusions about the interest and importance of these developments. They might well expand into new and extensive fields of work. It appeared that the systems designed by these three groups were completely different and largely incompatible. Time would show which system gave the better results. It was clear, however, that before long transatlantic baselines would be required, and Jodrell Bank had therefore held preliminary discussions with the Canadian group. They had proposed that if a system could be tested successfully, it should then be tried between Jodrell bank and a telescope in Canada, probably that at Penticton, in order to obtain the resolving power of 1/100 arc sec which would probably be required. Dr. Galt of Penticton had responded very warmly to these preliminary discussions. The R.R.E. group were now invited to co-operate with Jodrell Bank in completing the British system and operating it across transatlantic baselines. Dr. Hey said that for his part he would welcome an opportunity to participate in this exciting programme, and he was almost certain that his establishment would agree. He therefore welcomed the proposals in general terms, but wished to know in considerably greater detail precisely what R.R.E. were being asked to provide or to do. After considerable discussion it was proposed that they might help with four specific problems.

1. Provision of tape-recorder(s). Jodrell Bank proposed to test their system with one new Ampex recorder, and probably a second poorer instrument which was already on the station. They were applying for funds to order a second one. The system would be greatly improved if R.R.E. could provide or borrow one, preferably two, further wide band recorders. It was quite clear that for the operation of a transatlantic baseline three recorders would be required, in order that the instrument at the distant observatory need not be shipped back every time tapes were to be replayed and correlated.

2. Jodrell Bank would welcome advice and know-how with operation and transport of frequency standards at the two observatories.

3. Jodrell Bank would welcome any help which could be provided towards the transport of instruments, observers, and tapes to a transatlantic observatory.

After further discussion it was agreed that when a tape-recording system had been shown to work at all it would be tested betwen Malvern and Jodrell Bank, perhaps repeating observations made with the radio link interferometer. This would give experience in the use of the system to the Malvern

* British Columbia.
† The U.S. National Radio Observatory in West Virginia.
‡ But only because I knew we had lost! (See p. 64.)

team, who might then provide one or two observers for work at more distant sites.

4. Information on the figure of the Earth, and hence of the length and direction of the baselines.

Transatlantic collaboration

The stimulus for these developments came from the discussions which I had in the Soviet Union in the summer of 1963 and as far as we were aware the first idea that independent tape-recordings should be used was promulgated in the memorandum which we sent to the Academy of Sciences in the autumn of that year. Ironically this collaborative research eventually involved us with the Canadians and Americans and not with the U.S.S.R. Shklovsky had visited the National Radio Observatory at Green Bank in W. Virginia. Whether it was his transmission of the ideas in our memorandum to the Americans which started them on their tape-recording system and eventually led them to the joint measurements with the Crimea telescope I do not know. Perhaps some future historian in international scientific relations will be able to establish this interesting point. At least some of our own students who migrated to W. Virginia during those years had the impression that this was the case.

Eventually, neither the Americans nor ourselves were first with such a system. That honour belongs to the Canadians and as will be seen from the record of our meeting with R.R.E. in September 1966 we were already aware that our leadership in interferometric resolving powers was nearing its end. Indeed, at the moment of Palmer's return from the U.S.S.R. in 1964 I received a surprising letter from one of my old students V. A. Hughes* writing from Queen's University, Ontario.

Hughes to Lovell 30 June 1964: The Electrical Engineering Department is quite strong on the systems side and on signal analysis and Chisholm† with one of his students is developing a broad-band post detector correlation equipment for use with aerials with very long baselines. It is proposed that eventually the signals at two sites will be recorded on tape and then brought together to perform the analysis. The equipment has reached the stage where it is possible to correlate a noise signal and in the very near future will be tried out on two aerials over a baseline of about 10 miles. . . . When the system is completed we intend to use it with the radio telescope at present under construction at Algonquin Park and due to be completed in 1966, together with some of the other largish telescopes in Canada.

It was nearly three years later when the following telegram was

* V. A. Hughes was then Associate Professor at Queen's.

† R. M. Chisholm who made such important contributions to this development died in October 1968.

handed to me. Galt* to Lovell 23 May 1967: 'Long baseline interferometer Penticton to Algonquin using tape recorders and atomic clocks shows fringes for 3C 273 at 448 MHz resolution of one fiftieth second of arc.' Over the next few months the details of this remarkable measurement emerged from the published accounts—and furthermore revealed the narrow gap which separated the Americans and the Canadians in this drive to get beyond our own two million wavelength baseline measurements. On 28 March 1967 the Canadians communicated a short note to the journal *Science*† describing the successful test of this new technique. They had obtained tape-recorders previously used for television and had employed rubidium frequency standards for timing and the individual control of the local oscillators. Although this initial test was only over a short baseline of 200 metres between the 46 m Algonquin telescope and a small 10 m instrument, the experiment was completely successful.

On 5 June 1967 the Americans communicated a similar note to *Science*‡ describing the system they had developed and tested successfully over short baselines. Although similar in principle the Americans used a different recording system. The video signal was sampled at a rate of 720 kHz and the resulting bits were recorded on a high speed digital magnetic tape drive. The tapes from the two sites were then correlated in a digital computer. Neither of these initial experiments added significantly to the existing data on the angular sizes of the radio sources. They did, however, demonstrate clearly that systems were operating which would soon mark another revolution in our understanding of the structure of the radio galaxies and quasars.

On 1 July 1967 the Canadian authors published in *Nature*§ their account of the first use of the tape-recording interferometer over a longer baseline than any hitherto used. Between Algonquin and Penticton the distance is 3074 km and on the frequency of 448 MHz the effective baseline was 4·6 million wavelengths. With this arrangement they had, at that time, succeeded in measuring the fringes from two sources—

* Dr. J. A. Galt, worked at Jodrell Bank from September 1957, on leave from the Dominion Observatory in Ottawa. Early in 1959 he returned to Penticton B.C., where the new Canadian radio telescope was installed.

† N. W. Broten, T. H. Legg, J. L. Locke, C. W. McLeish, R. S. Richards of the National Research Council, Ottawa; R. M. Chisholm of Queen's University, Kingston; H. P. Gush and J. L. Yen of the University of Toronto, and J. A. Galt of the Dominion Astrophysical Observatory in Penticton; in *Science*, **156**, 1592, 1967 (23 June).

‡ C. Bare, B. G. Clark, K. I. Kellermann of the National Radio Astronomy Observatory, W. Virginia; M. H. Cohen, University of California, San Diego, and D. L. Jauncey of Cornell; in *Science*, **157**, 189, 1967 (14 July).

§ Broten, etc., *Nature*, **215**, 38, 1967.

3C 273B and 3C 345—and were able to conclude that the angular diameters were less than 0·02 seconds of arc at that frequency. It was these measurements to which Galt's telegram to me referred.

The American groups soon equalled this baseline. In June–July 1967* transcontinental interferometers were established between N.R.A.O. at Green Bank, the Haystack Antenna near Boston, the Arecibo telescope in Puerto Rico, and Hat Creek in California. Between N.R.A.O. and Haystack the distance is 845 km. At a wavelength of 18 cm the baseline is 4·7 million wavelengths. Of the 28 sources studied 11 were unresolved implying angular diameters less than 0·01 seconds of arc. The baseline between N.R.A.O. and Hat Creek was 19·5 million wavelengths and a new lower limit of angular size of 0·005 seconds of arc was set for some sources.

In January and February of 1968 the American groups worked with Swedish radio astronomers to establish a baseline of 6319 km between Green Bank and Onsala in Sweden—a baseline of 35 million wavelengths at 18 cm and of 105 million wavelengths at 6 cm.† A number of radio sources were still unresolved indicating angular diameters of less than 0·001 seconds of arc. Then, in September and October 1969 the international measurements with the Crimean installations were made‡ —not with the deep space telescope as I had envisaged in 1963 but with Moiseev's telescope near Simeiz (for which Palmer had expressed a preference, see p. 58). This distance of 8035 km established a baseline of 286 million wavelengths at 2·8 cm—far greater than we had hoped to establish six years earlier. Of the 12 sources studied 3C 345 was found to be unresolved with a diameter less than 0·0004 seconds of arc and there were several with diameters less than 0·0006 seconds of arc.

Meanwhile our own system was slowly emerging. It was soon clear that we could only be third in the attempt to establish multi-million wavelength transatlantic baseline interferometers, and months before we were ready to operate both the Canadian and American measurements had demonstrated that several radio sources remained unresolved over such baselines. We were, therefore, searching for some other unique contribution which we might eventually make with the tape-recording system. In September of 1967 Gold§ visited us and we discussed the possibility

* *Astrophys. J.*, **153**, 705, 1968.

† *Astrophys. J.*, **153**, L 209, 1968.

‡ *Astr. Zu.*, **47**, 784, 1970. (In translation *Soviet Astronomy A.J.*, **14**, 627, 1971)—a short note with 13 authors from the Crimean Astrophysical Observatory, the Institute of Cosmic Research and the Physical Institute in the U.S.S.R.; the Chalmers Institute in Sweden; and the N.R.A.O., Cornell and Cal. Tech. groups in the U.S.A.

§ Professor T. Gold, F.R.S., the director of the Centre for Radiophysics and Space Research of Cornell University.

of making observations between Jodrell Bank and Arecibo—the new feature being that two larger telescopes than hitherto employed would be used and hence the measurements could be extended to fainter sources.

According to the timetable which I sent him in November of that year we hoped that Anderson would be able to take the tape-recorder, clock, and other associated equipment to Puerto Rico in the spring of 1968. However, we were already moving into the difficult period of scheduling telescope time because of the initial phases of the conversion to the Mk IA. The system first had to be tested between the Mk I and Mk III telescopes, and it was March of 1969 before the first link with Arecibo was tested. Even this attempt was unsuccessful. All that one can do during the actual operation is to make sure that each system is working satisfactorily. Both Jodrell and Arecibo were, independently, working normally at the time of the measurements, but when the Arecibo tapes were returned to Jodrell for correlation, no fringes could be found. The reason was never established unambiguously, but was probably associated with phase stability problems. Fortunately the next attempt in November–December of 1969 was completely successful and fringes were obtained from 11 out of the 12 sources investigated. The separation was 6400 km—a baseline of 13 million wavelengths at 610 MHz giving a resolution of about 0·003 seconds of arc. More measurements between Arecibo and Jodrell Bank were made in March of 1970—one of the last uses of the Mk I before it went out of use for the final stage of its conversion to the Mk IA. The Mk I telescope was no longer available in 1971 but a further successful series of measurements was made with the tape-recording interferometer in March 1971 between the Mk II radio telescope at Jodrell Bank and the telescope at Onsala in Sweden—a baseline of 5 million wavelengths at a frequency of 1667 MHz.

Although our own system did not operate successfully over multi-million wavelength baselines until 1969, it is a pleasure to record that the Mk I telescope was itself involved in the first transatlantic interferometer experiment in 1968.

Galt to Lovell 18 December 1967: You will recall that we discussed the possibilities of a long baseline interferometer experiment between Jodrell Bank and Canada during my last visit there. At that time neither the Jodrell nor the Canadian apparatus had been tried, so we made no definite arrangement. The situation has changed now that we have had several successful observing sessions between the Penticton and Algonquin Observatories. . . . I would like now, to continue our discussions and explore the possibility of taking our apparatus to Jodrell sometime this Summer (1968) to make transatlantic measurements.

These arrangements with our Canadian colleagues went extremely smoothly and between 27 and 29 June 1968 the measurements were made between the Mk I and both Algonquin and Penticton. The distance to Algonquin is 5127 km, and to Penticton 6833 km which on the frequency of 408 MHz gave baselines of 7 million and 9·4 million wavelengths respectively. On 19 August I received a telegram; McKinley* to Lovell: 'Preliminary reduction of the data taken during the long baseline interferometer experiment ... has shown fringes from eleven sources.'

A fine example of the records produced by this experiment is shown in Fig. 9 of the source 3C 286, taken from the published account of this work.†

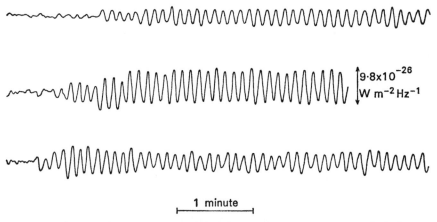

9·8x10^{-26}
W m^{-2}Hz^{-1}

|—— 1 minute ——|

9. Fringes observed from the source 3C 286 during observations between Jodrell and Algonquin. The projected baseline is $3·88 \times 10^6 \lambda$ and the position angle 132°. The noise level in the absence of fringes is shown at the start of each record. The relative delay between the two signals in the correlators producing the upper and lower traces differs by $\pm 0·15 \mu$ s from that in the centre trace.

* D. W. R. McKinley of the National Research Council, Ottawa.

† N. W. Broten, R. W. Clarke, T. H. Legg, J. L. Locke, J. A. Galt, J. L. Yen, and R. M. Chisholm, Mon. Not. R. astr. Soc., **146**, 313, 1969.

7

The Mark II, III, and IV radio telescopes

The developments related in the last chapter, concerning our work on the tape-recording interferometer, occupied only a part of the activities of the interferometer group after the achievement of the 2 million wavelength baseline with Malvern in 1966. At that time Anderson, with one or two research students, began to devote most of his attention to the longer baselines, while Palmer with the larger fraction of the group concentrated on other developments which were coming to fruition in 1966. In order to complete the story of the Mark I interferometric activities it is necessary to return to the situation in 1960–1 which stimulated this other line of development.

The Mark II and Mark IV radio telescopes

It is not my intention here to describe in any detail the history and use of the Mk II telescope but some reference to it is desirable in order to provide the background for the story of the Mark III telescope which was specifically designed for the interferometric programme in association with the Mark I. Undeterred by the terrible financial problems which still surrounded the Mk I telescope, but inspired by its successful operations I had in 1959 agreed with Martin Ryle to place a joint paper before the D.S.I.R. on 'The future of radio astronomy in Great Britain'. This was considered by the Astronomy sub-committee in the spring of 1960. I wanted to build a telescope larger than the Mark I and had many informal conversations with Husband* about the possibility of doing so. As a result of this paper, the D.S.I.R. agreed to make a grant of £10 000 for a 'design study to be made of a steerable radio telescope of significantly greater size than the 250 ft instrument now in operation at Jodrell Bank.'

By 'significantly' I had in mind a telescope vastly bigger than the Mark I—extending for a 1000 ft or more. Husband was bothered about

* Dr. H. C. Husband, C.B.E., senior partner in the firm of consulting engineers of Sheffield and London, and designer of the Mk I telescope. See *The Story of Jodrell Bank*.

the height problem particularly in relation to differential wind effects with height. We therefore conceived the idea of a telescope elliptical in shape with the height limited to manageable dimensions of 300 to 600 ft but with the horizontal dimensions extended to a 1000 ft or more. In general terms Husband thought the idea was practicable because he would be able to anchor the telescope along its horizontal axis through a series of bearings.*

In line with these general concepts and after discussions of my plans with the Board of Visitors to the Mark I, and informally in D.S.I.R., I made an application on 19 December 1960 for £173 228 for the 'Construction of a prototype radio telescope'. After referring to the design study grant mentioned above my application continued:

As part of this general development application is now made for a capital sum for the construction of a prototype telescope with elliptical aperture of 125 ft × 83 ft 4 in. . . . . The instrument is required for two reasons—(a) The very large telescope which is the subject of the main design study will have to be of radically different design from the 250 ft instrument if it is to be feasible from the engineering and financial aspect. Messrs. Husband & Co. have designed the prototype telescope in accordance with our general ideas of the future instrument. It is believed that the major instrument could be satisfactorily achieved by scaling up the prototype design as submitted in the engineers' report and drawing . . . accompanying this application. The construction and operation of this prototype would provide a sound practical experience for the testing of these basic ideas before embarking on the major instrument. . . . (b) It is unlikely that the design studies of the large telescope can be completed in time for a formal application before 1962 or 1963 and hence, financial considerations apart, it will be many years before the instrument will be operative. In the meantime new research facilities are required at Jodrell Bank both because the 250 ft telescope is overloaded with work, and for the pursuit of new lines of research.

In this latter connection I specifically indicated that since the new telescope would be more accurately figured than the Mark I it could absorb nearly all the hydrogen line work, and that it would be used to determine accurate positions of radio sources by the lunar occultation technique using it as an interferometer with the Mk I. I said that these two programmes would absorb two-thirds of the time of the new telescope and that the remaining one-third would be used 'on new research programmes, particularly on frequencies above 1420 MHz'. Rainford† countersigned this application on behalf of the University

---

* This project became known as the Mk IV and this explains the gap in our well-known nomenclature between the Mk III and the Mk V which puzzles even members of my staff who are unacquainted with this detail of our history.

† Dr. R. A. Rainford, O.B.E., Bursar of the University of Manchester until his retirement in 1971. See *The Story of Jodrell Bank*.

and in my covering letter to him with a degree of optimism which, today, I find remarkable I wrote: 'The D.S.I.R. also know informally about the project and a sum of approximately £250 000 is itemized in their astronomy papers to cover this and another less expensive item which we are not yet ready to pursue.' This other 'less-expensive item' was the Mark III telescope—a less accurate and transportable version of the Mark II, with a reflector of the same size and shape but of wire mesh. My application for this—for £109 850—followed two months later, on 21 February 1961.

The figure of £250 000 which I had talked about in D.S.I.R. a few months earlier had already become £283 000 and although I was dismayed that immediate authority was not given I suppose, in retrospect, that it is understandable that the grant-giving bodies should exercise caution—the memories of Mark I were still too fresh. By early March of 1961 Jolliffe,* who was then dealing with our affairs in D.S.I.R., said that there was unease in Treasury circles on two counts: first, that I was proposing to build two telescopes simultaneously, and second, that if they agreed to the 'prototype Mark II' they might be held to be making a commitment to build the telescope for which that was a prototype and which they knew would cost between £10 and £20 million.

So, once more, the arguments and lobbyings began. To begin with I rewrote the Mark II application, no longer calling it a 'prototype' and placing the main emphasis on the researches rather than the step towards the larger telescope. By April I was being asked to define priorities.

Lovell to D.S.I.R. 7 April 1961: ... it is our desire that priority should be given to the application for the Mark II telescope.... it is of course, unfortunate that a decision has to be made between the two, and this does not imply that we have any intention of abandoning the programme for which the Mark III transportable telescope is intended. ... it is our intention to withdraw the existing application for £109 850 for the transportable telescope plus associated electronic equipment. We intend to replace this with an application for a minor grant in the sum of £80 750 for the transportable telescope itself. ... in the meantime we are reconsidering what can be done about the test gear and electronic equipment which added the other £29 100 to the original application.

The D.S.I.R. then began to worry that we might be asking for considerable sums for the operation and maintenance of the Mark II. With the help of Rainford we reassured them that the U.G.C. would be asked for the rather small additional sums. All then seemed set fair and by mid-June I was hinting confidentially to Husband that 'good news might be

* Christopher Jolliffe, C.B.E., his long and helpful association with our affairs ceased when he became the director of the University Science and Technology Division on the formation of the Science Research Council in 1965.

released about the Mark II telescope by about mid-July'. A press conference was arranged by the D.S.I.R. Press Officer for 18 July. On 7 July it was cancelled and on 17 July:

Melville* to Lovell: I have to write to warn you of an inevitable delay in making a grant for your new radio telescope. . . . As you can imagine the present economic situation of the country is having repercussions on the finances of the Department. . . . The Research Council . . . therefore decided to suspend all the major new grants that are imminent. . . . I need hardly say that they did this with the greatest reluctance. . . . We should know where we stand by October; I hope that by then we shall be able to go forward with at least some of these major grants, including yours.

Undaunted, I replied to Melville suggesting that we should take this opportunity of proceeding more quickly with the design of the Mk IV. News of the difficulties which the Americans were encountering at Sugar Grove (W. Virginia—their proposed 600 ft telescope) increased our own enthusiasm for the Mk IV, and on 20 July I wrote to Jolliffe 'If we could get on with the design study soon we ought to be able to bring forward plans in a year or so for an instrument which would be without parallel anywhere on Earth.' We never did make progress with that telescope. The bureaucratic tie-up of major scientific grants (for which the Mark I troubles were held to be directly responsible) began to tip the scales heavily against such far-reaching projects unless unquestionable support could be obtained from the scientific community involved. But this I could no longer obtain. Even the Jodrell community of scientists began to want a diversity of equipment. A fashionable theoretical game amongst those who were using the Mark I, but who had no experience of the practical issues of getting such an instrument was to begin a conversation by saying 'Given £ thousand, what would we really like to build?' The answers were never unanimous, neither was the fallacy in the premise of the conversation ever understood. The Mark I was an act of faith, so was the Mark IV, but only the first vision was realized.†

Fortunately the delay in the Mark II was short lived. At its December

* Sir Harry Melville, K.C.B., F.R.S., who was Secretary of the D.S.I.R., during the major financial crisis on the Mk I, became the Chairman of the Science Research Council on its formation in 1965. In 1967 he resigned on his appointment as Principal of Queen Mary College University of London.

† In 1963 when I complained to Academician Keldysh in Moscow that I wanted £20 million for this telescope but was unable to get the money he said, after a moment's thought during which he presumably transferred pounds sterling to roubles, 'That is only a small per cent of the annual budget which I have at my disposal as President of the Soviet Academy of Sciences. Come to the Soviet Union and we will build it for you.' I thanked him sincerely but said that I was too devoted an Englishman to accept his kind offer.

meeting the Research Council of the D.S.I.R. decided to give priority to the Mark II over other competing projects which were before the Council. On 18 December 1961 the Chief Press Officer* of the D.S.I.R. wrote to me about the arrangements for the release of the news—'I am sure you could not wish for a better piece of news at Christmas than this'—and on 21 December the papers carried the information that we were to build another telescope. It is not my intention to describe this telescope further except to remark that the delay had one important consequence in that we changed our mind about the site for the instrument. We decided to remove the 218 ft transit telescope and build the Mk II in its place using part of the neighbouring 'Park Royal' building as a control room. This was a decision for which I was responsible but which I now regret. We should not have demolished the transit telescope. In any case the construction of the Mark II proceeded without incident and by the summer of 1964 we had the telescope under test, since when it has played a notable part in our researches, either used singly, or as an interferometer with the Mark I.†

The Mark III radio telescope

I have described above how the application for £109 850 to build the Mark III was made on 21 February 1961, just two months after I had asked for a larger sum to build the Mark II. Today it is difficult for my students to understand why the Mark II and III telescopes have elliptical paraboloids as reflectors instead of a circular 100 ft paraboloid, which would be the conventional equivalent in gain and collecting area. The explanation in the case of the Mark II was given above. Since the Mark III had to be cheap and transportable it seemed clear that it would be advisable to follow closely the Mark II concept, thereby easing Husband's design effort and facilitating eventual transportation of the instrument.

Now, a decade later, I regret both of these decisions. To begin with, the Mark IV has remained an idea, and the concept of the Mark II as its prototype (which it was fundamentally intended to be) has had no value. Of course, Husband designed the reinforced concrete turntable arrangement for the Mark II before he built the first Goonhilly telescope (although the latter was in use before the Mark II)—and it is possible that he and his design team were beneficially influenced by the concept

---

* Norman Manners who succeeded to this office on the retirement of Col. Hingston who figured so prominently in the public relations of the Mk I.

† Technical details of the Mark II have been given in the Appendix to my book *Our Present Knowledge of the Universe*, Manchester University Press, 1967; see also B. Lovell, *Nature*, **203**, 11, 1964.

of the Mark II. Secondly, the Mark III has not yet been moved from the site where it was built—the urgent need to use the instrument at varying baselines has not yet materialized because we have been able to continue the arrangements for the use of one of the R.R.E. telescopes at Defford. The Mark II and III have both been splendid instruments, but the difficulties of feeding the elliptical aperture and the need for symmetry in many of the researches which developed subsequently means that they are frequently used with feeds which illuminate only the circular 83 ft section of the paraboloid.

Whereas the Mark II was designed to extend our operations to shorter wavelengths (which it has done admirably), in addition to the prototype concept, the Mark III was proposed as a less accurate telescope forming an essential part of the interferometric facilities for the measurement of angular diameters. In 1960–1, Hanbury Brown and Palmer were just completing the survey of the angular sizes of 384 radio sources at baselines up to 32 000 wavelengths. This work has been described in Chapters 4 and 5. Apart from the need for an extension of the baseline with which the subsequent account has been primarily concerned, two other outstanding requirements had emerged. Many of the sources had been shown to be of complex structure and it was necessary to study the angular distribution in a variety of position angles—necessitating a continuous tracking of the sources across the sky. Secondly, the size of the existing remote aerials used with the Mark I were setting the limit to the number of sources which could be studied, and those 384 were becoming a small fraction of the total number catalogued by the Cambridge surveys.

These requirements were summed up in the application to the D.S.I.R.

... it is now proposed to carry out a third survey of the angular sizes of radio sources. It is clear that we need to identify more radio sources and to establish their luminosity function, brightness temperatures and linear size. The results of the second survey at Jodrell show that this cannot be done by measurements with a fixed resolving power in one orientation. It is necessary to have an easily variable resolving power and to examine sources in different position angles. It is therefore proposed that a new interferometer be constructed using the 250 ft steerable paraboloid as one element and a transportable, steerable 100 ft paraboloid as the other element.* The variable resolving power and orientation will be obtained firstly by observing sources over a wide range of hour angles and secondly by using several frequencies in the range 178–1000 mc/s. Coarse variations in resolving power and orientations will be obtained

* Although '100 ft paraboloid' appears in the application it is quite clear that we intended from the beginning that this should be the 125 ft × 83 ft 4 in copy of the Mark II—the '100 ft' being the equivalent in collecting area.

by changing the site of the 100 ft paraboloid. It is not visualized that this would be carried out more than about three times in a survey which might last for a period of about three or four years. With the proposed combination of aerials it is estimated that detailed information on the distribution of intensity across about 650 sources could be provided. This would represent an adequate sample of the radio source population.

In the application which was sent to D.S.I.R. on 21 February 1961, the cost of this telescope was estimated to be £80 750 (this included the design fees) and we asked for an additional £29 100 for electronic equipment. This figure also included a sum of £4500 which was then the estimate of one removal to another site. Husband estimated that it would take 3 months to design in detail and 12 months to build.

I thought that it would increase our chances of getting this grant if the application was made by Hanbury Brown since I had already applied formally for the Mark II. Therefore, he signed the application form, but since it was countersigned by Rainford and me I cannot imagine, in retrospect, how we thought this would influence the D.S.I.R. It did not, and this manoeuvre was to boomerang unpleasantly. In any case, from my comments on the Mark II it will be clear that by April of 1961 the great optimism with which the year opened was already evaporating. On 7 April, as related on p. 71, I defined our priority as between the Mark II and III and in a further manoeuvre I asked that the Mark III grant should be modified to remove the £29 100 for the electronic equipment. The idea was that this would leave the application for £80 750 and since this was less than £100 000 D.S.I.R. would then treat it as a 'minor grant' and it might then come into a more favourable part of the division of their resources.

This idea was stillborn from the start:

D.S.I.R. to Lovell 17 April 1961: . . . there is little point in submitting a completely new application for the same project. . . . You mention a total of £80 750, which I believe includes no allowance for a contingency item. Without an estimate from the Ministry of Works* for the cost of contribution it would seem reasonable to include a 10% contingency (£7500). . . . The amendment of Professor Hanbury Brown's application from a 'major' to a 'minor' application has come about because of an arbitrary level set by the Research Grants Committee. This is an internal D.S.I.R. matter.

The letter also said that the Department would need an assurance that I could find the balance of £29 100 elsewhere. 'As the survey could not

---

* After the Mark I troubles the Treasury now insisted that grants for all major constructional items should be handled by an 'agent'. In my subsequent experience this proved to be a most costly, inefficient, and delaying arrangement.

be carried out without the equipment, an assurance on this aspect is a necessity before the application can be approved.'

From the brief account of the fate of the Mark II application given above it will be clear that in the next few months the financial situation deteriorated rapidly and it soon became a question as to whether we would get even one of the telescopes. For the rest of that year, apart from another upward revision of the estimates of cost in October, all efforts were centred on the Mark II to which we had given priority. Indeed, the Mark II grant awarded in December 1961 was in such competition with other projects (for example in nuclear physics) that further efforts to get another large amount of money from D.S.I.R. in the immediate future were clearly destined to fail and I began to seek other avenues for financing the Mark III. It was not that the D.S.I.R. staff or their advisers were lacking in enthusiasm for the Mark III. I was then on the Astronomy sub-committee, and early in 1962 there was a discussion in which the hope was expressed that the sub-committee might find £40 000 from their allocation for astronomy. Since the official indication was that the D.S.I.R. would have no possibility of finding the money for the whole project until 1964 or 1965 I tried to interest people elsewhere in joining in with the project.

In March 1962 I set out for Oregon where I had to give the Condon Lectures.* On the way I went to Harvard and asked Leo Goldberg† and Fred Whipple‡ for their advice. Acting on their view that a frontal approach to the National Science Foundation would not succeed, I followed their advice and from Corvallis, Oregon, on 4 April 1962 I wrote to J. L. Pawsey who had recently been appointed the director of the National Radio Observatory at Green Bank, W. Virginia. Whipple and Goldberg had told me that Pawsey made the possibility of international collaboration a condition of his appointment and I was full of optimism again:

Lovell to Pawsey 4 April 1962: This letter is a strictly personal and provisional approach concerning an issue of possible future collaboration between Green Bank and ourselves. It arises out of discussions which I had a few weeks ago with Goldberg and Whipple, who suggested that I might write to you. About two years ago I applied to the D.S.I.R. for the sum of £120 000 so that we

* *The Impact of Modern Astronomy on the Problems of the Origins of Life and the Cosmos,* Oregon State 1963.

† Professor Leo Goldberg, of Harvard College Observatory, succeeded Professor D. H. Menzel as Director and in 1971 was appointed the Director of the Kitt Peak Observatory, Arizona.

‡ Professor Fred L. Whipple, Chairman of the Department of Astronomy, Harvard University 1949–56 and Director of the Smithsonian Institution Astrophysical Observatory since 1955. See also Ch. 13.

could construct a telescope which is known as our Mark III. This instrument is rather a crude copy of our Mark II in that it consists of an elliptical bowl with an effective aperture of 100 feet designed so that it can be easily transported around the countryside. It is intended for use with our 250 foot reflector for the continuation of the programme of Palmer's group on the measurements of the angular diameter and structure of the radio sources. I expect you heard Palmer's account of his work last year at Berkeley in which he discussed his work on the 384 radio sources with intensity greater than 12 flux units, which he has studied with baselines out to 61 000 wavelengths involving a separation of 72 miles from Jodrell Bank. The analysis of this work which now has been sent in for publication indicates very strongly that an extension to a lower range of flux units is essential if we are to study in any detail the sources which have redshifts greater than about 0·5 c. All the calculations have been done and it seems certain that with the proposed combination of the Mark III and the Mark I we could extend our present limit of 0·5 c to 0·65 c, involving measurements on about 600 sources.* Unfortunately since the D.S.I.R. has just given me £250 000 for the Mark II it does not have enough money to finance the Mark III although all the advisers regard the programme with enthusiasm. In fact I have been informed that there is no chance of getting the complete sum until 1964 or 1965. On the other hand, I think that it is conceivable that about one half of the sum required might be made available this year. From the above you will have appreciated that the purpose of this letter is to make a provisional inquiry as to whether you would consider the possibility of joining with us in a 50 per cent finance and operation of this important programme. I am, of course, writing on a strictly personal basis, though the individual mainly concerned in D.S.I.R. knows that I might be speaking to you about this. I would have to establish it on a formal basis if your initial reactions were in any way favourable.

Alas, within a few days I received in Portland, Oregon, a reply containing tragic news from Dr. F. H. Nicoll, Pawsey's brother-in-law. It was written from the Washington Hospital Centre and postmarked Princeton 9 April. '. . . he is in good spirits and feeling well but unable to use his left arm or leg. The doctors are hopeful that this condition will not persist and Joe maintains his usual optimism. He sends you his best regards.' Unfortunately the doctors' hopes did not materialize. Joe Pawsey had been struck with a mortal illness on the eve of assuming his appointment at Green Bank.†

* The certainty at that time that we could steadily increase our penetration to greater redshifts is interesting. This was before the impact of the discovery of the quasars.

† The death of J. L. Pawsey at the age of 54 was a tragic loss. He was one of the great pioneers of radio astronomy and, indeed, was almost wholly responsible for the high international status which Australia had achieved in the subject at that time. He had recently decided to accept the appointment at Green Bank and was struck down only a few weeks after his arrival in the U.S.A. For details of Pawsey and his career see my memoir of him 'Joseph Lade Pawsey 1908–1962' in *Biographical Memoirs of Fellows of the Royal Society*, **10**, 229, 1964.

W. L. Francis* had been my main contact in D.S.I.R. with whom I had discussed these possibilities of international collaboration to get the Mark III under way. When I returned home I wrote to him:

Lovell to Francis 17 April 1962: I don't know if you have heard the bad news about Pawsey. Just after he arrived in the United States two or three weeks ago to take up his job at Green Bank he had a severe stroke, and is now lying in hospital in Washington in a bad condition. It seems, therefore, that there would inevitably be a considerable delay before we could make any progress in setting up a joint programme with them on our Mark III. . . . At the last meeting of our sub-committee hope was expressed that D.S.I.R. might be able to find about £40 000 towards the project from their current resources. Are you in a position to indicate to me informally if you think that this would be a possibility if I could devise some means of financing the rest of the project by raising the money at this end? I am anxious to keep the project under our sole control in Great Britain provided we can do this without delaying it for so long that other people will carry out the experiment first.

I cannot remember how I intended raising the balance. Looking back now I think it would have been a difficult task after our experiences with the Mark I appeal—and so did Francis who also had another angle on the idea. He replied on 19 April confirming that subject to approval at higher levels the Astronomy sub-committee was prepared to allocate £40 000 but

We could not make a partial grant of this kind without receiving a revised estimate of the total cost and a firm guarantee that you could secure the balance from other sources. You are the best judge of your likely success in seeking outside support for £60 000 or more but after the earlier experience with Mark I you might find this difficult. Such an experience would also expose D.S.I.R. and H.M.G. to public criticism for parsimony, which we would take harshly in view of the substantial grants we have recently made.

Francis then went on leave and it was several weeks more before I could talk to him about the problem. The outcome of these further discussions was that I would make another approach in America—this time jointly to the Office of Naval Research and the National Aeronautics and Space Administration. The draft letter and detailed explanatory material was agreed with Francis, and on 28 May I sent the letter and documents to Dr. A. Shostak† of the Electronics Board of the Office of

---

* Dr. W. L. Francis, C.B.E., was then the Director of the Grants and Information Divisions of the D.S.I.R. On the formation of the Science Research Council in 1965 he was appointed the Secretary of that new body.

† My association with Dr. Shostak was of long standing and for many years he had arranged for us to receive a helpful grant for our researches from the U.S. Office of Naval Research. It was this grant (of about 20 000 dollars per annum) to which, during a period of student unrest in the University of Manchester in 1969, the students

Naval Research in Washington, to Dr. Homer Newall, the Director of the Office of Space Science of N.A.S.A., and to Dr. Robert Seamans, Associate Administrator of N.A.S.A. The essence of the approach was that there should be a three-pronged finance with D.S.I.R., O.N.R. and N.A.S.A. each contributing £40 000.

On 20 July 1962 Dr. Seamans replied in a kind and understanding letter but '. . . it has been N.A.S.A. practice not to support ground-based astronomy, except in those areas of research which specifically support the space flight programme. As interesting as your proposed research is, it does not seem to meet this criterion, nor does the instrumentation appear to hold sufficient promise for satellite probe tracking to meet our support on this basis.'

I have no record or memory of a reply from Shostak—he may well have spoken to me at a meeting—but as I remarked to Francis after getting Seaman's reply I had little hope from O.N.R. since they had just cancelled the 600 ft Sugar Grove radio telescope. By September I was suggesting to Husband that since we were making no progress in financing the steerable telescope we might build the bowl only and fix it on the ground so that it pointed towards the zenith. We would then make pre-set adjustments in elevation by tilting up to about 45° by using guy ropes or hydraulic jacks. Husband thought we could do this for the £40 000 which D.S.I.R. might make available.

However, before we could make any progress with this new idea I heard from Jolliffe in mid-October that the Research Grants Committee would shortly draw up a new list of priorities for the major outstanding projects and that in his opinion the Mark III might be high on the list. Then on 25 October Jolliffe wrote to confirm that the R.G.C. had recommended that the project should proceed. By the end of that year we were already thoroughly enmeshed in the Ministry of Works machine. Amongst other matters their representatives wanted to increase the estimate to include £12 500 for concrete aprons and access roads and to increase our allowance of £4500 for the first move by ten times! This was merely the beginning of an appalling series of costly delays both on the Mark II and III occasioned by this insistence on using agents without experience in these matters*—and moreover who demanded about

---

objected and claimed that we were 'helping the U.S. in the Vietnam War'. I advised the Vice-Chancellor (Sir William Mansfield Cooper) to inform the students that I acknowledged the help from the O.N.R. and to explain to them that I used it to finance my collaborative research on a nearby star with my colleagues in the Soviet Union. He passed on this information with marked success.

* We were particularly incensed that M.P.B.W. placed an *architect* in charge of the highly technical matters concerning the Mark II and III telescopes.

10 per cent for themselves, yet would assume no responsibility for the fallibility of their own estimates.

The insistence of M.P.B.W. that it would cost £40 000 to move the telescope instead of the £4500 estimated by Husband proved a serious stumbling block to the acceptance of a revised figure by D.S.I.R. Finally, on 25 January 1963 we compromised on a figure of £10 000, and in my letter to Jolliffe of 28 January dealing with this I was able to make the following additional comment:

> There is one other point relevant to this [the cost of moving] which I made during the discussion. In the original application it was visualized that the site of the Mark III might be changed by not more than three times in 3 or 4 years. At the time of this document in early 1961 the electronic techniques then available indicated that we might be able to study 650 sources with this telescope combined with the Mark I. Successive improvements in our techniques now indicate that the number of sources which we should be able to study with this combination has increased to between 5000 and 10 000. Because of this the change of site of the telescope will now be much more infrequent, and thus although our estimate for Item 15 has gone up, the overall cost of the programme for a period of five years will not have increased because it is likely that we would only require to shift it once during the first five years from the site on which it is erected initially.*

By mid February M.P.B.W. had transmitted to the D.S.I.R. their version of the specification and estimated the cost of the telescope alone as £115 000 (to be compared with the £80 000 for which I had asked a year earlier; £15 000 of the increase was the agency fee). It was on this basis that, at last, after more tribulations, I learnt in August that Treasury approval had been obtained. The official announcement of the grant by D.S.I.R. was dated 27 September 1963.†

One of the delays was an unfortunate backfire of my plan that Hanbury Brown should sign the grant application. By this time, as related in Chapter 3 his interests were beginning to be concentrated in Australia. Naturally, D.S.I.R. in full knowledge of Hanbury Brown's movements and intentions could scarcely make the grant to him. In April after much telephoning and correspondence between Sir Harry Melville and the Vice-Chancellor, D.S.I.R. were satisfied that the research could proceed as planned in Hanbury Brown's absence and so the grant was formally transferred to my name. My idea about the signature had been useless from every point of view. Two and a half years had already elapsed since the proposal to build the telescope was

---

* The Mark III has still (1972) not been shifted from the original site.

† A separate grant of £23 850 was subsequently made, direct to the University (that is excluding M.P.B.W. interests) for the electronic equipment.

made. Further disillusionment lay ahead. Husband thought that the telescope would not be in use until May 1965 even if he received instructions to proceed immediately (that is by 1 October 1963). The M.P.B.W. maintained that even these dates 'seem to us to be rather idealistic'.

Throughout all these protracted discussions we had never been sufficiently heartened to take positive steps about finding an actual site on which to build the telescope. Now, in the autumn of 1963 the question suddenly became urgent. I asked Palmer to set out his current thoughts on this question and on 1 November he produced a memorandum outlining the salient points which should guide us in this matter. The starting point was that the recent measurements had clearly indicated that the very small sources of high surface temperature might be comprising a group containing the most remote sources. They represented about one quarter of the 384 sources then studied and the immediate objective in using the Mark III was to select a similar fraction of the additional 1500 sources which would become readily available for measurements with the Mark I and III interferometer. Taking account of the spectra of the sources, the preamplifiers available, and the performance of the telescopes, a frequency of 408 MHz was considered to be optimum. The existing measurements on the 384 sources indicated that a maximum baseline of the order of 20 000 λ would be appropriate—in other words a separation in the region of 10 miles from the Mark I.

In the event a fortunate combination of circumstances directed our attention to a site somewhat more distant. Beyond ten miles south-west of Jodrell Bank there is a large area of the open country of the Cheshire Plain. Sixteen miles south-west lay Haughton Hall the home of a friend—Mr. Geoffrey Dean. His land extended a few miles towards Jodrell Bank and near the village of Wardle at its north-east extremity was bounded by the Shropshire Union Canal. He was reclaiming a large area from its wartime usage as an airfield. Palmer had already worked a small remote aerial on one of the crumbling runways, fifteen miles from the Mark I, and by mid November I had established with Geoffrey Dean that he would be willing in principle, for us to use this as a site for the Mark III. On 29 November I took Husband to see him and the proposed site. From Husband's end there were no problems, the existing access from the nearby road was adequate. The next stage was to persuade Rainford to enter into the necessary financial arrangements for the lease of the land and once more good fortune followed, because it transpired that Rainford and Geoffrey Dean were old golfing companions. Everything was settled with remarkable speed on an issue which might well have caused us trouble in a less favourable personal climate.

By mid April 1964 the tenders showed reasonable accord with the estimates. The contract was let on 21 April. Although we had gained a month on the original target date for letting this contract the M.P.B.W. target for completion of 31 July 1965 was distressingly far ahead. By the end of 1964 only the azimuth track and centre pivot were in position at Wardle and according to the M.P.B.W. progress report the 'manufacture of the bowl and superstructure is proceeding very slowly and is now six weeks behind programme'. By May of 1965 the target for completion had gone back to the end of August 1965 and the steelwork manufacture and erection was still six weeks behind programme. A month later it was eight weeks behind. By mid August when we should have been using the telescope, progress was 'still unsatisfactory in spite of heavy pressure being brought to bear on the Contractors' (M.P.B.W. 15.8.65). The target date went back another month. On 23 August I received a letter from S.R.C. which caused me to lose my temper.

You will recall from the last meeting of the Project Committee ... that 31 August still remained a feasible date for completion. There has since been a substantial further slippage in the programme. ... The view of the M.P.B.W. is that even with favourable circumstances the end of September is a more likely date and that in the event the second half of October might prove a more realistic assessment.

On 24 August 1965 I replied to the S.R.C. letter in a manner which did little except release my own feelings.

24 August 1965: I visited the site with Husband on August 19 and hence am well acquainted with the miserable situation related in your letter. Once again I have to place on record my strongest disapproval of the way in which this job has been handled by the Ministry. As soon as one penetrates into the day to day details. ... a condition of inexcusable inefficiency is revealed. ... The situation on the Mark III has, of course, been made even worse by the fact that —— may be good at building aeroplanes, but they have negligible experience of outdoor constructional work of this nature.

In fact my strictures about the agency arrangements were widely shared. They led eventually to a meeting at which the University and the S.R.C. interviewed competitive agencies for our future telescopes then envisaged. I cannot say that the decisions then taken marked any significant easing of the problem. That, however, is part of another story. The Mark III continued to drag on interminably. By mid November it was to be the end of that month. At last by the end of 1965 the telescope was finished apart from minor works—or so we imagined. Unfortunately the testing of the telescope revealed a series of problems with the hydraulic drives and by the spring of 1966 a new completion date of September was in mind. In fact, we accepted the telescope on 14 April 1966 subject to various reservations the most important of which concerned the azimuth drive. Gradually we infiltrated the radio equipment on to the telescope and made the arrangements for its remote control by radio link from Jodrell Bank. On 11 October 1966 I was able to send to Husband a full report on the radio performance of the telescope: 'You will be pleased to observe from the conclusion in paragraph 6 that we are now satisfied that the position indicator and the performance of the reflector meet our requirements and fall well within the specification.'*

The Mark I and Mark III as an interferometer

When the Mark III was proposed early in 1961 it was conceived as a cheap telescope, quick to build, to enable us to extend within a year or so the important series of angular diameter measurements. Unfortunately, for the reasons related above it was another five to six years before we could use it in any systematic manner. But for our good fortune in gaining access to the Malvern telescopes the delay would have been a disastrous matter for our researches. As it was, during those intervening years the outlook on angular diameter measurements had changed dramatically—and so had the progress of astronomy to which those measurements were related. Quasars had been discovered and the need for resolving powers of fractions of a second of arc had monopolized our attention. When the Mk III eventually became usable in 1966 we were achieving our 2 million wavelength baseline measurements with Malvern. The choice of the baseline for the Mark III turned out, in the event, to be more fortunate than we had imagined. Almost the entire field of structure measurements smaller than ten seconds of arc remained relatively untapped. Further, the bowl of the telescope turned out to be better than we had specified and by using it on 21 cm we were able to increase still further our range of spectral investigations and resolving powers.

The technique of observing simultaneously on more than one frequency was developed by Rowson and the mass of data which rapidly accumulated stimulated the development of digital computer techniques for the on-line analysis of these data. By the time the Mark I was placed in the hands of the engineers in August 1970 the Mark I+III combination had worked for 4300 hours, measuring during this time the angular diameters of 300 radio sources and the detailed structures of

* Technical details of the Mark III telescope may be found in the Appendix to my book *Our Present Knowledge of the Universe*, Manchester University Press, 1967, and H. P. Palmer, and B. Rowson, *Nature*, **217**, 21, 1968.

200 of these. In addition the Mark I had been linked to Malvern to fill in more data and with this combination another 700 hours of measurements had been made involving the study of 50 sources and the detailed structures of 20.

One of the first uses of the Mark I+III combination was to take up the important cosmological problem of the variation of angular size with redshift. One of Palmer's students G. K. Miley made measurements on 72 quasars using a frequency of 408 MHz giving a baseline of 32 000 wavelengths. The redshifts of these quasars were known and 65 of them gave recognizable fringe patterns. Unfortunately, at that time, the initial phases of the engineering work for the conversion of the 250 ft telescope to the Mark IA were involving urgent repairs to the azimuth railway track and the telescope could be used only at a fixed azimuth. Under these circumstances Miley could not measure the fringe visibility at different hour angles in order to obtain some idea of the structure of the sources. Nevertheless, his preliminary results gave strong support to the view that the mean angular dimensions of these quasars decreased with increasing redshift.*

Miley, following many of our students, migrated to America and at the National Radio Astronomy Observatory at Green Bank he was able

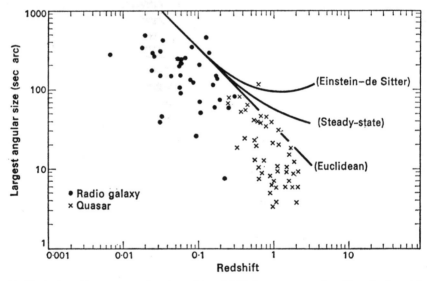

10. Variation of largest angular size with redshift for quasars and radio galaxies with spectral indices steeper than −0·5 and no low-frequency cut-off. The solid lines are the theoretical predictions for a source of linear size 400 kpc on various cosmological models (see Ch. 9).

* G. K. Miley, *Nature*, **218**, 933, 1968.

to continue these investigations. A further 79 quasars were investigated and structures were derived for 36 of them. An analysis of all the available information from this work, coupled with various other measurements subsequently published, enabled Miley to confirm the previous indication that there was a significant decrease in the angular size as the redshift increased and furthermore that there was a continuity in this relationship as between radio galaxies and quasars. Fig. 10, taken from Miley's published account* of this work, indicates the important cosmological deductions which may be possible when the various influences of spectra type and evolution on the sizes are better understood. In this diagram the largest angular size of the source has been plotted against the redshift. The solid lines in the diagram show the relation to be expected on three simple cosmological models. Since the radio measurements refer to a three-dimensional object projected on to the sky, one would naturally expect a scatter of these observed sizes below the theoretical curves even if all the sources had the same physical dimensions.

* G. K. Miley, *Mon. Not. R. astr. Soc.*, **152**, 477, 1971.

8

The impact of computer technology

When the Mark II telescope came into use in the summer of 1964 it was controlled on-line by a digital computer—a Ferranti Argus 100. It was certainly the first telescope to be controlled in this way and probably the first major machine of any type. This courageous and brilliantly successful step was largely due to the insistence of J. G. Davies. His unparalleled knowledge of such devices and their application to these tasks steadily wore down my own innate mistrust and conservatism. Within a few years almost the entire field of our operations was transformed by the application of digital techniques. The computer control of various parameters of the actual receivers, and the on-line and off-line analysis of the data became common features of the researches. Measurements of incredible speed and complexity became possible, particularly with the associated development of the digital spectrometer designed by J. E. B. Ponsonby and constructed by him and his associates.

The first application of digital techniques to the interferometer measurements

During the spring and summer of 1968 a research student, P. K. Wraith*, working with Palmer's group, made the first successful series of simultaneous measurements on two frequencies—150 MHz and 408 MHz—between the Mark I and Mark III telescopes. This also marked the first application of digital techniques to the interferometric measurements. These developments by Wraith were carried out as part of the research for his Ph.D. degree. The analysis of the analogue charts to compute the variation of fringe frequency with hour angle was becoming a laborious and time-consuming occupation—particularly with an interferometer working simultaneously on more than one frequency. Wraith tackled the problem of transforming the output of the interferometer into a digital form suitable for computer processing.

* P. K. Wraith, now of the Department of Medicine, Royal Infirmary, Edinburgh, was at Jodrell Bank from 1965 until 1971.

The first stage in the process was to feed the video output of the interferometer into a voltage-to-frequency converter (VFC). This electronic unit converted the input signal to a square wave with a frequency proportional to the input voltage. This output was counted electronically so that a binary number was obtained proportional to the input of the VFC integrated over a specified interval. However, there are a number of technical difficulties, such as the requirement that the separate interferometer channels need separate digitization and therefore have independent gains. To overcome this and other associated problems Wraith used a cyclic counter which had been devised for other purposes by C. G. T. Haslam.* In this system the output signal for the VFC is counted in one of four divider chains selected by an electronic switch. This switch was actuated by the reference waveform controlling the phase sensitive detector so that each counter was actuated for one quarter period of the reference cycle. A suitable combination of the numbers stored in these four binary counters was then fed on-line to an Argus 400 computer.

The complex procedures used in this system have been described in detail by Wraith†. The final inteferometer output was on punched tape giving the following information: the fringe amplitude averaged over the preceding minute; its standard deviation; and a pair of fringe phases, being the average phase over the first and second half of the integration period. The tape also contained information about the time, source hour angle and the telescope following errors. Further processing of this tape, to include various corrections, led eventually to a print-out of all the relevant data of fringe amplitude as a function of hour angle for the two separate frequencies of 151 and 408 MHz which Wraith used in his observations.

In 9 weeks of observation between the Mark I and Mark III telescopes Wraith produced 15 miles of punched tape. Reliable structure information was obtained for 48 sources. 69 per cent were doubles; 10 per cent had an unresolved core surrounded by an extended halo, and 21 per cent appeared as single components. In the latter class 12 per cent were unresolved at either frequency. An important result was that there were only small differences in the structure on the two frequencies. By comparing the data from known quasars and unidentified sources in his survey, Wraith concluded tentatively that the unidentified sources lay

* C. G. T. Haslam came to Jodrell Bank as a research assistant in 1958 after graduating in the University of Manchester. He was awarded his Ph.D. degree in 1962 and remained at Jodrell Bank until 1970 when he was appointed to the Staff of the Max-Planck Institute for Radioastronomy to work with the new German radio telescope near Bonn.

† P. K. Wraith, Ph.D. Thesis, University of Manchester 1970.

at the same order of distances and had the same range of radio luminosities as the quasars, but were optically subluminous. In the second part
of 1968 Wraith's technique was used to make similar measurements on
40 quasars between the Mark I and Mark III telescopes on frequencies
of 408 and 610 MHz. The full implication of these results still awaits
further study in relation to subsequent measurements on further samples
of sources at 408 and 1420 MHz, but at this moment the combination
of all these data indicates that the unidentified sources may be remote
radio galaxies.

The digital autocorrelation spectrometer

Although Wraith's logging process described above was the first application of digital techniques to the interferometer measurements, it has
been quickly superseded by a technical advance of great significance to
such measurements. Frequently, it is found that an instrument designed
for one purpose has unexpected and important influences on other
researches. The digital spectrometer has been an instrument of this kind.

By 1964 the Mark I telescope had been used extensively to study the
1420 MHz spectral line from neutral hydrogen, both in the galaxy and
in extragalactic nebulae. Measurements were also beginning on the
recently discovered spectral lines, in the 1600 to 1700 MHz region,
from the OH radical. With the receiving techniques then available
much of this line radiation was extremely weak, and long integration
times were required to record it. Broadly speaking the technique of
investigation was to drive the telescope on the selected region of sky and
then obtain the spectrum (that is intensity as a function of frequency) by
scanning the receiver in frequency. This implied using, for example, an
overall bandwidth of say 200 kHz and using filters in the receiver so that
the intensity in a whole series of bands of perhaps 5 kHz could be
selected. By this means only a small fraction of the spectrum could be
observed at any one time and since integration times of an hour or more
might be necessary these researches were extraordinarily time consum-
ing and laborious.

A technique which would enable the whole spectrum to be recorded
simultaneously and integrated for as long a time as necessary was clearly
highly desirable. Apart from the investigations of the H and OH lines
an instrument of this type would facilitate the search for new radio
spectral lines which, at that time had been predicted theoretically, but
not observed.* An experimental spectrum analyser with 21 channels
and a switch rate of 75 kHz capable of handling signals in a bandwidth

---

* This situation changed dramatically in the following years. The development of
instruments of the type discussed here coupled with the evolution of radio astronomi-

up to 150 kHz, had been built by Dr. S. Weinreb of the Massachusetts Institute of Technology. This analyser was successfully used in measurements on the Zeeman splitting of the 21 cm H line, and it was then evident that more comprehensive instruments of this type would revolutionize the approach to many current radio astronomical problems. During the summer of 1964 R. D. Davies,* who was in charge of our line measurements at Jodrell Bank, pressed me for money to develop a much more comprehensive and faster analyser which he believed would be possible with the new digital techniques then developing at Jodrell Bank. He wanted nearly £8000 to purchase the various components and since there was no hope of my finding this money in our budget I suggested to the Executive Secretary of the Royal Society that this was the type of new scientific instrument in which the Committee of the Paul Instrument Fund might be interested. With his encouragement I arranged for Davies to make a formal application, which he did on 21 August 1964. The application was successful and on 21 December 1964 Davies learnt that the Paul Instrument Fund Committee had agreed to a grant of £7800 for the development and construction of the digital spectrometer. At that time a switch rate of 4 MHz had already been achieved by preliminary work at Jodrell Bank and at this rate a bandwidth of 1 MHz could be analysed fully. It was planned to use 120 channels. In fact, even further advances were made during the development of this instrument and with the help of a further grant of £4132 from the Paul Instrument Fund in the spring of 1967 a far more ambitious spectrometer eventually came into use in 1968.

The spectrometer has 256 channels and is capable of processing signals in a bandwidth up to 5 MHz. The filters are such that any one of 8 possible bandwidths from 5 MHz to 39 kHz may be selected. The successful development and construction of this instrument depended heavily on J. E. B. Ponsonby† and L. Pointon‡. The spectrometer does

cal techniques in the millimetre range soon led to the discovery of many spectral lines from complex molecules, in the interstellar medium. The paper 'Interstellar Molecules and Dense Clouds' by D. M. Rank, C. H. Townes, and W. J. Welch in *Science*, **174**, 1083, 1971, gives a list of 22 molecules discovered to the end of 1971. Three more were discovered in 1972.

* Dr. R. D. Davies, came to Jodrell Bank from Adelaide as an Assistant Lecturer in 1953. His distinguished contributions to the study of neutral hydrogen in the galaxy and in extragalactic nebulae form a large part of the Jodrell Bank researches which are not described in detail in this book. He was appointed to a Readership in 1967.

† J. E. B. Ponsonby came to Jodrell Bank from Imperial College as a research student in 1960. He was appointed to the Staff as Assistant Lecturer in 1963 and promoted Lecturer in 1966. His work on the radar detection of the planet Venus is described in detail in Ch. 14.

‡ L. Pointon, a Senior Experimental Officer at Jodrell Bank.

not merely measure the power spectrum $P(f)$ of a signal $x(t)$ by the direct use of filters and individual detectors. It determines the time delay autocorrelation function $R(\tau)$

$$R(\tau) = \int x(t).x(t-\tau)dt$$

which is related to the power spectrum through the Fourier transform

$$P(f) = \int R(\tau).\cos(2\pi f\tau).d\tau$$

where f is the frequency and τ the time delay. By using a digital store to delay the signal a number of alternative overall bandwidths may be explored according to a chosen clock rate. The incoming signal is 'clipped' so that only polarity information is preserved before the autocorrelation function is taken, thus only a binary or 'one-bit' representation of the signal is required.

The spectrometer was designed to work on-line with an Argus 400 computer which carries out the Fourier transform and associated computations. A detailed description of the instrument and its mode of operation has been published* and a typical H line spectrum directly reproduced on a plotting-table is shown in Fig. 11.

The application of the digital spectrometer to the interferometer programme

Reference has been made to the work of Wraith who digitized the output of the interferometer and effectively replaced the pen-and-ink charts by an output on punched tape. Although this greatly facilitated the handling of large amounts of data it did not, in principle, affect the fundamental possibilities of the interferometer technique. However, a revolution soon followed the introduction of the digital spectrometer devised for quite a different purpose. In all the work with two linked telescopes so far described the fringe patterns have concerned one radio source. Now the beamwidth of the telescopes in much of this work has been of the order of a degree, thus covering an area of sky far greater than necessary for the observation of a source with angular diameter measured in seconds, or minutes of arc at the most. Furthermore, within this beam coverage other radio sources exist, but remained unobserved because the delays and other parameters of the equipment must be adjusted for one source only. For example with the Jodrell Bank–Defford baseline using a bandwidth of 1 MHz, the usable part of the aerial beam for a given equalization of the path lengths is, perhaps, only about 9 minutes of arc. In other words, with two sources $\frac{1}{2}$ degree apart in the beam simultaneously, only one could be observed. The

* R. D. Davies, J. E. B. Ponsonby, L. Pointon, and G. de Jager, *Nature*, **222**, 933, 1969.

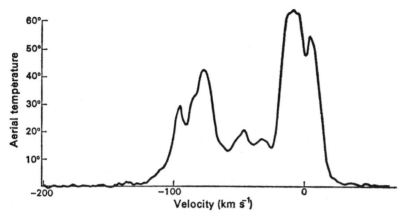

11. A typical hydrogen line (λ21 cm) spectrum directly reproduced on a plotting table by the 256 channel autocorrelation spectrometer using the Mk I telescope in the direction galactic longitude $l = 92.76°$ and galactic latitude $b = +4.43°$.

concept of using several delay channels in order to observe more than one source simultaneously remained little more than a theoretical possibility until the advent of the digital spectrometer.

Such an arrangement would, in any case, be unworkable with the existing systems because of phase instability between the two telescopes. The fluctuations in the phase arise partly from lack of phase coherence in the links between the telescopes and partly because of ionospheric fluctuations—although the latter are not serious for frequencies above 1000 MHz. Because no long baseline interferometer system working on frequencies lower than 1000 MHz has achieved the necessary phase stability, these measurements have concentrated essentially on the investigation of the amplitudes of the interference fringes and the variation with hour angle, and with different separations of the telescopes. These existing techniques have enabled the angular sizes of radio sources to be measured and, in simple cases, the structures to be deduced by the process of model fitting, provided they have not been too complex. For example, it has been possible to decide whether the Fourier transform is equivalent to a single or double source; but in the latter case, where the components are unequal, it has not been possible, without the phase information, to decide whether the stronger component lies east or west of the weaker. In recent years considerable attention has been given to the development of systems which, at least for high frequencies where the ionospheric irregularities become unimportant, would enable long baseline interferometers to achieve phase stability.

In 1969 R. J. Peckham, a research student in Palmer's group, achieved an elegant partial solution of this problem, which enabled him to use the digital correlator to observe more than one source in the beam simultaneously. Essentially, the procedure depends on always having one source in the beam of the telescopes which is unresolved at the baseline in use. This source is then used as a continuous phase calibration for the signals from any other source or sources which may be in the beam simultaneously. The technique deals both with system and ionospheric phase fluctuations even at low frequencies—up to source separations of $\frac{1}{2}$ degree at least. Given one source as a phase reference in this way, there are further problems in separating and displaying the fringes from this source, and others in the beam. Because of the displacement in the sky the delay will be different for the path lengths from the sources. The new correlator had already been used in the tape-recording interferometer experiment to search in delay. Thus resolution could be obtained in this way because the fringes from two sources simultaneously in the beam generally occur in two different delay channels. This gives resolution in one direction. Resolution in another co-ordinate is obtained by simultaneous analysis of fringe frequency. Eventually an on-line map of the area of sky in the telescope beams is produced at chosen intervals, the two co-ordinates being delay and fringe frequency.

Peckham tested this system between the Mark II radio telescope and the telescope at Defford on a frequency of 408 MHz. The geometrical mean of the half-power beamwidths of these two telescopes at that frequency is 2 deg 6 min. The digital correlator was used to cross-correlate (not auto-correlate) two different signals. A range of delay channels was chosen to sample 1 degree within this beam at all values of hour angle and declination—a requirement of 8 microseconds in overall delay. Sampling was carried out using 30 correlator channels, with a separation of 0·4 microseconds between adjacent channels. The Argus 400 computer was programmed to log the correlator results after each 5 second integration period. The computer stored these data for 64 integrations (320 seconds) and then performed a Fourier analysis to plot out a point source map of the area of sky covered by 1 degree of the telescope beam at those 320 second intervals. An example of the point source map appearing automatically on the plotting table is shown in Fig. 12.

The success of this technique immediately led to entirely new possibilities for the interferometric observations. In particular Peckham demonstrated that the system could be used (a) to obtain fringe visibility curves for pairs of sources separated by as little as 17 minutes of arc. Sources with this order of separation in the sky are far too close to be studied by existing interferometer techniques. (b) to carry out long integrations on a weak source which lay within 1 degree of an intense

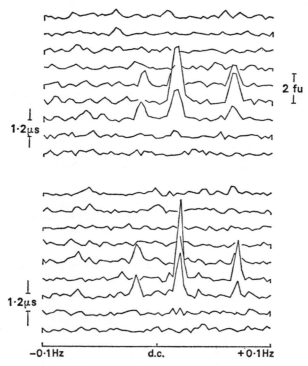

1·2μs

2 fu

1·2μs

−0·1Hz   d.c.   +0·1Hz

12. Two successive correlator plots of 320 seconds integration showing the radio source map of the sky covered by 1 degree of the telescope beam using the Mk II to Defford interferometer. The prominent sources in this map are 3C 303·1 and 3C 305·1.

source used for phase reference. A sensitivity level of 1 flux unit was achieved after 5 minutes integration; but over the Defford–Jodrell baseline integrations could be carried out for approximately 30 minutes and in some cases for 53 minutes—leading to sensitivities of 0·1 flux unit.

At this stage Palmer suggested that the technique could be used to study the region of sky surrounding strong radio sources. The immediate areas near the stronger radio sources are largely unsurveyed because of the effect of side-lobe responses in nearly all survey techniques. By using a long baseline interferometer many of the strong sources can be resolved sufficiently to reduce the side-lobe responses by orders of magnitude, while still leaving sufficient intensity in the main beam to serve as a phase reference source. For example, on a frequency of 408 MHz the Cassiopeia source has an intensity of 5920 flux units, whereas the flux density seen by the Jodrell Mark II to Mark III interferometer is only 2 fu. Cygnus reduces from 4790 to 100 fu; Taurus from

1246 to 2 fu and Virgo from 543 to 10 fu. The use of the shorter baseline to Mark III increases the effective sampling area by 25 times to 30 square degrees—and in this case the actual area which could be surveyed was limited not by the fringe frequency/delay diagram, but by the beamwidth of the telescopes—about 2 degrees.

In one week of August 1971 the Mark II–III combination was used with the digital spectrometer to survey the regions around those 4 strong radio sources. Peckham discovered one new source near Cassiopeia of 0·6 flux units, 2 near Cygnus and 4 near Taurus. No new source was found near Virgo. Fig. 13 shows the records of one of the new sources (2003+38) near Cygnus; with an intensity of 1·1 fu and displaced by +05 min in right ascension and by +3 deg in declination from the strong Cygnus source. The most remarkable result came from the investigation of the Taurus region. The interferometer reduced the effective flux from the Taurus-A source from 1246 to 2 flux units and the correlator plots of 6 successive 320 second integrations are shown in

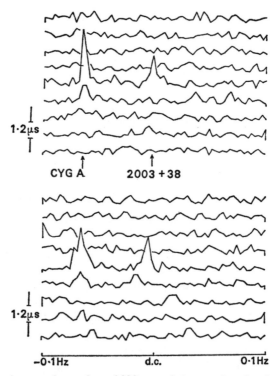

13. Two successive correlator plots of 320 seconds integration showing the use of the Mk II–Mk III interferometer to study the region of sky in the vicinity of a strong radio source. In this example Cygnus-A is partially resolved, the effective flux density being only about 1/50 of the intensity of the source.

Fig. 14. The records indicate the existence of at least 3 and probably
4 weak sources. One of these PKS 0531+19 was already catalogued.
This was used as a phase reference to make a 37 minute integration of
the region with the result shown in Fig. 15. These two strip distributions in adjacent delay channels reveal clearly 3 other sources in addition to PKS 0531+19. These 4 sources near Taurus-A all lie on a straight
line passing through the reference source PKS 0531+19. The probability is less than one part in 1000 of this occurring by chance and it
remains to be seen whether such results may indicate a physically connected chain of radio galaxies.*

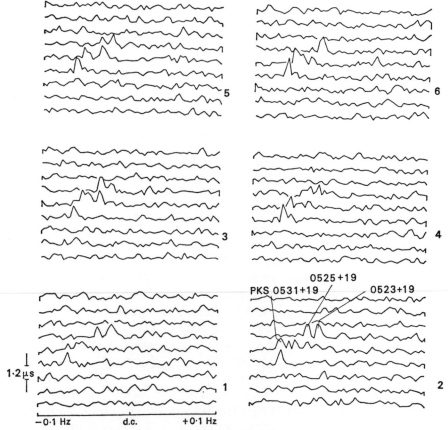

14. Six successive correlator plots each of 320 second integration using the Mk II–
Mk III interferometer to study the region of sky in the vicinity of the Taurus-A radio
source. Taurus-A is nearly completely resolved. (2 fu effective, instead of the actual
1246 fu intensity of the source on 408 MHz).

* Details of these measurements were subsequently published by R. J. Peckham and
H. P. Palmer, *Nature Phys. Sci.*, **240**, 76, 1972.

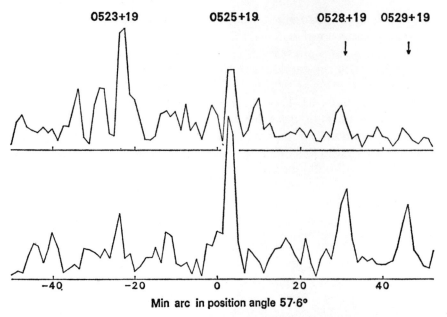

0523+19   0525+19   0528+19   0529+19

Min arc in position angle 57·6°

15. A 37 minute integration of the region of sky near Taurus-A shown in Fig. 14, using the source PKS 0531+19 as the phase reference. These are two strip distributions in adjacent delay channels, and show clearly the two sources 0525+19 (1·5 fu) and 0523+19 (1·0 fu) marked in Fig. 14 and two other sources (each of 0·65 fu) not evident in Fig. 14.

A new survey technique

The success of the correlator applied to the interferometer led Peckham to a new survey concept using long baseline interferometers. The system, known as 'the end of baseline survey' is a drift survey making use of the fact that for sources passing through the same hour angle as the baseline the fringe frequency reduces to zero. This exploits the large range of delay available with the digital correlator. The telescopes are used at fixed positions, so that without using the fringe speed machine or delay tracking, a broad strip of sky can be observed for sources down to the resolution limit and sensitivity of the system.

An opportunity to test this new idea arose during a Jodrell–Defford experiment in November 1970 when there was a temporary breakdown in the Defford telescope tracking system. The telescopes were fixed at the appropriate azimuth and elevation to observe a strip of sky centred on declination 33 deg from right ascension 01^{h}30 to 10^{h}00. The choice was made because a suitable calibration source 3C48 would be observed at the start of the observations. On two successive nights 10 sources, already catalogued, were successfully identified on the records.

Subsequently, the end of baseline survey technique was used to make a study of the percentage of sources smaller than a given diameter (in this case $\frac{1}{2}$ second of arc) as a function of intensity. One of the early uses of the Mk IA telescope in 1972 was in a refinement of the end of baseline technique to survey nearly a thousand sources, measured firstly between the Mk IA and Mk III and then between the Mk IA and Defford, in an attempt to make further progress with the investigation of this relationship between angular size and intensity.

9
Relevance to cosmology

Much of the work described so far in this book has been stimulated by the belief that the results could make an important contribution to our understanding of the nature of the universe. The theoretical investigation of the cosmological problem and the efforts to obtain decisive observational evidence in favour of particular models, has monopolized the attention of many astronomers of diverse interests during recent years. The great resurgence of interest arose when it became evident in the few years after 1950 that the radio telescopes had revealed extragalactic objects of unknown provenance—the radio galaxies —and that these were observable at much greater distances than the classical types of galaxy catalogued by Hubble and his associates. Until that time the possibility of reaching a decision between the multiplicity of theoretical world models depended on the large optical telescopes in their study of these normal galaxies, or the large clusters of them. Since 1950 many possibilities of, apparently, decisive observational tests have emerged from the use of the radio telescopes.

Of course, the universe as we understand it observationally today, is an entirely different concept from the universe in which the theorists of the early part of this century were interested. Although Herschel, in the early nineteenth century, formed a good qualitative idea of the shape of the Galaxy and speculated about 'island universes', the general belief was that the stars of the Milky Way defined the totality of the universe. Further, the Sun was believed to be near the centre of the system. Nebulae, although well observed, were all believed to be gaseous objects within the confines of the Milky Way. Any motions of the objects in the universe were believed to be small. In other words, an explanation had to be found for a static assembly of stars. On Newtonian theory the force of attraction between the stars would obey the inverse square law and no obvious basis existed for the theoretical explanation of such a universe, if the gravitational law applied throughout all space. The imponderable nature of the theoretical problem was well appreciated and near the turn of the century suggestions were made that, for the universe as a whole, the gravitational law should contain a term so that at large distances repulsive forces balanced Newtonian attraction and thereby led to a static universe.

Since this concept was at variance with all measurable and directly observable features it is hardly surprising that interest in cosmology waned in the early years of this century. It was revived abruptly in 1916 with Einstein's publication of the general theory of relativity. As regards the nature of space and gravitation, general relativity differed in vital respects from Newtonian concepts. In Newtonian theory space was absolute, and the gravitational attraction between two bodies of mass m_1, m_2 separated by a distance r was $G\frac{m_1 m_2}{r^2}$ where G is the constant of gravitation. In general relativity space is not absolute; the properties of space are a function of the bodies in the universe, with the gravitational force having equivalence to the deformation of space near a massive body. In Newtonian theory gravitational and inertial mass appeared as separate entities—the gravitational mass determining the force exerted on another body, while the inertial mass determined the resistance of a body to motion. Although they appear as separate entities in Newtonian theory, no measurement had ever been able to detect any difference between them. Indeed, already in 1890 refined measurements had shown that the gravitational and inertial mass of a body were the same to at least 1 part in 100 million. In general relativity the identity appears naturally because the gravitational forces are manifestations of the inertia of bodies in space, which is itself modified by the presence of these bodies.

In 1893 the Austrian physicist Mach had already rejected the Newtonian concept of absolute space. Mach argued that the behaviour of one body in the universe could only be considered in relation to all other bodies. For example, one can measure the angular velocity of the Earth either by observing its motion with respect to the stars, or dynamically by using a pendulum or gyroscope. The answer is the same, and hence the astronomically and dynamically determined frames of reference are the same. The conclusion appears to be that the local inertial frame of reference is determined by some average of the motion of the distant astronomical objects. In 1917 Einstein sought to give expression to this principle of Mach in the field equations of general relativity and thereby unify the gravitational and inertial fields. However, difficulties of reconciliation with Mach's principle arise because the equations express the influence but not the entire cause of inertia since the boundary conditions at infinity are not given. Einstein could not choose boundary conditions such that the inertial field was fully determined by the masses in the universe. He overcame the difficulty by introducing an additional term multiplying the inertial field terms in the equation—the Λ term, with dimensions of the inverse square of length. With this Λ term positive Einstein found a solution of the field equations with uniform density of matter, random velocities zero and with space so curved that it was unbounded but finite. Thereby the difficulties at infinity are abolished. Einstein also showed that with Λ positive there was no solution for empty space and hence he believed that Mach's principle was incorporated fully in the theory.

At that time, in 1917, the measured velocities in the universe were small compared with the velocity of light. The universe was believed to be a static entity and so Einstein, in turn, believed that his field equations represented the universe to a first approximation. But within months de Sitter found another solution of the Einstein equations for an empty universe, which was only static if empty of matter. Hypothetical test particles introduced into the universe would recede from each other with ever increasing velocity. Einstein's solution with Λ positive was not unique and did not satisfactorily incorporate Mach's principle in general relativity.

It is a remarkable feature of astronomical history that de Sitter's work which showed that there were solutions of the field equations predicting a non-static universe, was followed in a few years by the observational evidence for the expanding universe. The key to progress was a new method for measuring stellar distances. The parallax measurements, first successfully employed by Bessel in 1838, to measure the distance of the star 61 Cygni were limited by the techniques of the nineteenth century to stars no more distant than about ten light years. In 1912 Miss H. S. Leavitt of Harvard found that in certain variable stars (the Cepheid variables, after δ-Cephei, the prototype measured by Miss Leavitt) the period of variability was related to the absolute magnitude of the star. Fortunately there were sufficient variables of this type within the group of stars of known parallax to establish the quantitative relationship and hence a powerful new method of measuring distances emerged, provided that the period and apparent magnitude of a Cepheid could be determined.

The first major results of this new distance measuring technique were obtained by Shapley between 1916 and 1919 in his studies of the Cepheids in globular clusters—leading to his proof that the stars of the Milky Way were not symmetrically disposed with the Sun at the centre, but formed a flattened disc of the type now so familiar in contemporary photographs of spiral galaxies. Then, in 1926 Hubble published the results of his observations of the nebula M33, made with the 100 inch

Mount Wilson telescope. He had been able to resolve the nebula into stars, and from the measurements of the Cepheid variables he produced the first unambiguous proof that the nebula was an extragalactic star system remote from the Milky Way. In the same year he published the results on 400 of these extragalactic star systems. Then in 1929 he related the distance of these nebulae to the redshift measurements of Slipher and Humason.*

The phenomenon of the redshift in the spectra from the nebulae had been observed by Slipher in 1912. Interpreted as a doppler shift very high recessional speeds were indicated. By 1924 Slipher had measured 43 spectra, 38 showed the redshift effect. He noted that the fainter nebulae exhibited the greater redshifts, indicating recessional velocities of up to 1200 miles per second. When Hubble measured the distance of these nebulae and proved that they were extragalactic star systems the final step was taken to establish the generalized picture of the expanding universe as we understand it today. In 1929–and 1931 Hubble* published his results defining the relation between distance and velocity (redshift)—a linear relationship out to 100 million light years where the recessional velocities were 2000 miles per second. Subsequently the 200 inch Palomar telescope extended the linear law to 2000 million light years and that was the limit of penetration at the time of the discovery of the radio galaxies referred to earlier in this book.

Theoretical predictions

The dramatic observational proof of the non-static universe followed de Sitter's solution of the Einstein field equations which showed that the universe was only static if empty of matter. Soon afterwards the Russian scientist A. Friedmann investigated the theoretical solutions of the field equations in which the radius of curvature and mean density varied with time. The mathematical treatment is complex, but in 1934 solutions on the basis of Newtonian theory were discovered by McCrea and Milne, which yielded identical predictions to the solutions based on the field equations of general relativity—although the physical attributes of certain terms in the solution differ fundamentally. Further, although identical models are obtained for the variation of the radius of the uni-

---

* The striking nature of Hubble's results were known to many astronomers before they were published. The actual publication of his classic series of papers occurred as follows. On M33 in *Astrophys. J.*, **63**, 236, 1926 (in which we referred to similar results on the Andromeda nebula M31 but those details were not published for another 3 years—*Astrophys. J.*, **69**, 103, 1929). On 400 nebulae, in *Astrophys. J.*, **64**, 321, 1926. The first publication on the distance–redshift relation was in *Proc. nat. Acad. Sci. Am.*, **15**, 168, 1929; with the relation extended to a distance of 32 megaparsecs in the paper with M. L. Humason, *Astrophys. J.*, **74**, 43, 1931.

verse as a function of time, there are strict limits to the interpretation based on the Newtonian concepts.

Both the Newtonian and relativistic treatments predict identically the variation of the scale factor R as a function of time t in the form

$$\left(\frac{dR}{dt}\right)^2 = \frac{C}{R} - k + \Lambda R^2$$

where C is a constant including the gravitational constant and the density of the universe at some particular cosmic time. However, the constants k and Λ have different interpretations on the two theories. In the relativistic treatment Λ (the cosmological constant) is the term introduced by Einstein to overcome the difficulty that in his field equations the boundary conditions at infinity were not given as mentioned on page 100. In the Newtonian theory Λ is a consequence of the term referred to on page 99, suggested by Neumann and by Seeliger in 1895 and 1896 as a means of preserving a static Newtonian universe. In the relativistic treatment, k determines the curvature of 3-dimensional space at any one time; on the Newtonian theory it is the total energy (kinetic plus potential). Clearly there is, in principle, an infinite set of solutions for the variation of R with t, unless some evidence can be obtained to determine k and Λ. At present, although many claims are frequently made to the contrary, it may be remarked that there is no widespread agreement even as to the sign (positive or negative) of the constants k and Λ.

The simplest solution, and the theoretical prediction to which observational results are most often compared, is the case where both k and Λ are zero. This is commonly known as the Einstein–de Sitter model. The variation of the scale factor with time is shown in Fig. 16. In this universe R varies as $t^{\frac{2}{3}}$ and the rate of expansion tends to zero as R tends to infinity. If the Hubble time* is taken as 10^{10} years then the present age since the moment of infinite density at zero time is $6 \cdot 7 \times 10^9$ years and the present density is 2×10^{-29} g cm^{-3}.

In the models with k less than zero, and Λ zero, the variation of R with t is of the type shown in Fig. 17. In the beginning R varies as $t^{\frac{2}{3}}$ as

---

* The Hubble constant H is the relation between the velocity of recession cz and distance r, $(H = cz/r)$, determined from the redshift measurements on extragalactic nebulae. The inverse of the Hubble constant (H^{-1}) is referred to as the Hubble time. The value of 10^{10} years corresponds to a Hubble constant of 100 km s^{-1} Mpc^{-1}. Sandage has recently published $(Q.J.R.\ astr.\ Soc.,\ \mathbf{13},\ 282,\ 1972)$ results of a new determination of $H = 55$ km s^{-1} Mpc^{-1} which gives a Hubble time of $1 \cdot 77 \times 10^{10}$ years. The Hubble time is the time from the beginning of the expansion of the universe for zero value of the deceleration parameter q. For the more general relationship involving non-zero values of q see Fig. 20. In the case quoted here for the Einstein–de Sitter model, $q = +\frac{1}{2}$.

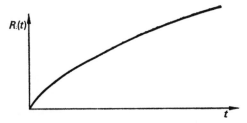

16. The variation of the scale factor of the Universe $R(t)$ with time (t) for $k = 0$ and $\Lambda = 0$. (The Einstein–de Sitter model.) The rate of expansion tends to zero as R tends to infinity.

in the Einstein–de Sitter universe, but then varies directly as t; so that the universe is ever expanding (in Newtonian theory the model is gravitationally unbound—the particles have excess kinetic energy). With k greater than zero, and Λ zero, R first increases with t and then decreases again to zero as shown in Fig. 18. There are no restrictions on an endless repetition of this process of alternate expansion and contraction.

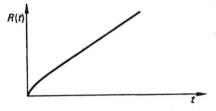

17. The variation of the scale factor of the Universe $R(t)$ with time (t) for $k < 0$ and $\Lambda = 0$. The rate of expansion decreases monotonically with increasing R but remains positive as R tends to infinity.

18. The variation of the scale factor of the universe $R(t)$ with time (t) for $k > 0$ and $\Lambda = 0$. The universe reaches a maximum size and then contracts as time increases.

If Λ is made either positive or negative, then, depending on the value of k, another family of model universes is predicted, including those which contract from infinity to a finite value of R and then expand again to infinity. The models studied by Eddington and Lemaître with Λ and k positive either start from a finite value of R at infinite past time, and then expand continuously; or (in the Lemaître model) start from infinitely small R at time zero, expand to a nearly stable finite value of R, and then move to a state of further expansion. Some years ago it was believed that the absorption lines in the quasar spectra were concentrated around $z = 1\cdot9$, and the intermediate nearly stable condition of the Lemaître model was invoked as a possible explanation. Lemaître visualized this model as beginning in a primeval atom which disintegrated at time zero. After an indeterminate but very long time the impetus of the initial explosion was exhausted and the primeval gas cloud settled to a nearly stable condition with a radius corresponding to the static solution of Einstein. With the forces of attraction and repulsion nearly in balance the Lemaître universe would necessarily spend a long time in this nearly stable condition before the eventual resumption of indefinite expansion.

For more than twenty years the major dispute has been concerned not so much with the values of Λ and k in the field equations of general relativity, but whether evolutionary theories of this general type are to be preferred to the steady-state theory. This theory, which emerged in 1948, was stimulated initially by two primary considerations. Firstly, the evolutionary theories already discussed have singular conditions in the past involving the problem of creation. For many values of Λ and k the singular condition is that the entire material of the universe must have been in a state of high density within a small volume at time zero. The existence and formation of this initial condensate is not an issue capable of discussion in physical terms. Indeed, for many, the creation of the universe in its initial state of high density is essentially a problem in metaphysics or theology. The steady-state theory sought to evade this problem of the initial condition, by bringing the problem of creation within the scope of contemporary inquiry.

Secondly, in 1948 there was a severe problem concerning the age of the Earth and the age of the universe. The age of the Earth was established from several independent lines of inquiry to be about 4·5 thousand million years. But, on the basis of the Hubble constant then current, the age of the universe (since the singular condition) on the evolutionary theories was only about 5 thousand million years. By no conceivable processes of condensation from the primeval gas into nebulae, stars, and planets could the Earth be considered to have emerged almost simultaneously in the very early history of the universe. In fact, several years after the publication of the steady-state theories it was found that the Cepheid variables in the extragalactic nebulae, on which the distance relation had been based, belonged to a different stellar population from those studied by Miss Leavitt and obeyed a different period–luminosity law. When the appropriate corrections were made the distance scales for the universe were doubled. The Hubble constant was thereby changed, giving an age from the singular condition of about 10 thousand million years. The conflict with the age of the Earth was thereby resolved, but the steady-state theory remained as an attractive proposition because of the first consideration.

In 1948 Bondi and Gold introduced the concept of the perfect cosmological principle—on the large scale the universe is unchanging, and possesses a high degree of uniformity both in space and time. Since the universe is expanding, the principle implies that new matter must be in process of continuous creation in order to maintain a constant density. As the nebulae move apart, so new ones are formed from the created matter, at a rate which is precisely sufficient to present an unchanging aspect to an observer anywhere in time and space.

The rate of creation of new matter demanded by the concept is, on the average, $3\rho H$ where ρ is the mean density of matter in the universe and H is Hubble's constant. Taking typical values for ρ and H this value is 3×10^{-46} g cm^{-3} s^{-1}. This rate is equivalent to the creation of one hydrogen atom per litre of space every 500 thousand million years. Since this rate of creation is beyond the limits of conceivable observation, the adherents of the theory maintain that there is no contradiction with the observations. Consideration of the consequences of these assumptions leads to the conclusion that the scale factor R of the expanding universe increases exponentially with the time t. This is the same scale factor as derived from the field equations of general relativity with $k = 0$ and Λ positive—the de Sitter model (not the Einstein–de Sitter model which has $k = 0$, $\Lambda = 0$), which had to be rejected when derived from the field equations of general relativity because of the implication that the equation could be satisfied only if space were empty of matter. The problem does not now arise in the steady-state model because the conservation of matter implicit in the relativistic field equations is rejected.

In the same period a different approach to the steady-state concept was proposed by Hoyle. His model was not based on the concept of the perfect cosmological principle, but he derived identical conclusions from suitable modifications of the field equations of general relativity in which the Λ term was abandoned. It should be mentioned that in recent years various modifications have been considered in the steady-state theory in an effort to retain the compatibility of this concept with apparently conflicting observational evidence.

The possibility of observational tests of cosmological theories

There are a number of observations which can be made on distant galaxies which, in principle at least, should make it possible to distinguish between the various cosmological theories outlined above. The assumption is made that the redshift is a doppler effect associated with the cosmological expansion of the universe. Measurements of the redshift of a distant galaxy will then give the velocity of recession of the galaxy. As we penetrate into space so we inevitably observe the universe as it was in a past epoch. If these observations can be made to a sufficient distance, that is 'past' time, then a critical parameter for distinguishing between the various theories should be obtainable.

For small redshifts, the classical doppler relation is accurate. That is the redshift $z = \delta\lambda/\lambda$ can be written as

$$z = \frac{\delta\lambda}{\lambda} = \frac{v}{c}$$

By using this relation Hubble first established the linear relationship between distance and redshift. However, as the value of z increases so that the indicated value of v approaches the velocity of light c, the classical relationship is no longer accurate and the relationship derived from the special theory of relativity must be used. This may be written conveniently in the form

$$1+z = \left(\frac{c+v}{c-v}\right)^{\frac{1}{2}}$$

This relationship between z and v is plotted in Fig. 19. Initially the relationship is linear, that is $z = v/c$ but already at $z = 0.5$ the relativistic effects are significant. For values of $z = 2.0$ (greater values are known for some quasars) $v = 0.8c$. As v approaches c the redshift tends to infinity.

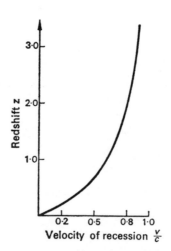

19. The relation between the redshift $z\left(= \dfrac{\delta\lambda}{\lambda}\right)$ and the velocity of recession v in units of the velocity of light $c = 1$. For small v the relationship is linear. As v approaches c, z tends to infinity according to the special theory of relativity.

If we were concerned with local effects, and the redshifts arose from straightforward movements of objects at high velocity, then these simple expressions could be used to measure distances and the elapsed time from the emission of the light to the moment of observation. However, in the observation of the redshift of the galaxies we are concerned with the expansion of the universe, and the interpretation of the measured values of z in terms of distances of the objects and the 'look back' time into the past history of the universe is dependent on the cosmological model assumed—particularly on the value of the quantities k and Λ. Indeed, by measuring the redshifts only of objects in the universe it would not be possible to make much progress in the investigation of the cosmological problem. We need to know the light time (look-back time) or the distance. Some of the approaches which have been made to these problems will be mentioned here.

(a) *The deceleration parameter*

The cosmological models with various values of k and Λ clearly make significantly different predictions about the change in the rate of expansion of the universe with time. If the velocities could be measured at various values of t, and sufficiently far back into the past history of the universe for evolutionary effects to be apparent, then it would be possible to derive the deceleration $-\ddot{R}$ (where $\ddot{R}$ is the second derivative of R with respect to t). It is customary to define a deceleration parameter q which is independent of the time t at which $R(t_0) = 1$ (t_0 being the present time from the singularity), and also such that it is dimensionless. The independence with respect to time is achieved by considering the quantity $-\ddot{R}/R$. Since this has dimensions (time)$^{-2}$ we multiply by $R^2/\dot{R}^2$ (which is equal to $1/H^2$, the inverse square of the Hubble constant). Then the dimensionless deceleration parameter q is defined as

$$q = -\frac{R\ddot{R}}{\dot{R}^2}$$

As an example of the immediate relevance of the determination of q to the cosmological problem it can readily be shown that for the Einstein–de Sitter universe in which Λ and k are zero

$$q = \frac{4\pi}{3}G\rho\, H^{-2} = \tfrac{1}{2}$$

where G is the gravitational constant and ρ the density of the universe. In the cases where $\Lambda = 0$ and $k > 0$ then $q > \tfrac{1}{2}$ and for $\Lambda = 0$, $k < 0$, then $0 < q < \tfrac{1}{2}$.

If q can be measured and the value of the Hubble constant H is

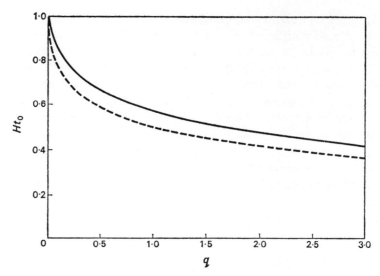

20. The relation between the time t_o from the singularity in the Universe, the Hubble constant H, and the deceleration parameter q, for a radiation-filled universe (——), and a matter-filled universe (- - - -), with $\Lambda = 0$.

known, then the time t_0 from the singularity in the universe can be determined. The relation between q and Ht_0 is shown in Fig. 20 for two types of universe in which Λ is zero.

The light travel time t_l in terms of time since the beginning of the expansion t_0, that is t_l/t_0 has been calculated by Sandage* for various values of q (with $\Lambda = 0$). Some illustrative examples are given in the following table:

| | | | | q | | | |
z	0	0·1	0·5	1·0	2·5	5·0	∞
0·02	0·0196	0·0232	0·0293	0·0340	0·0433	0·0533	0·1777
0·461	0·3155	0·3656	0·4337	0·4730	0·5258	0·5615	0·6756
0·50	0·3333	0·3856	0·4557	0·4952	0·5477	0·5826	0·6919
1·00	0·5000	0·5693	0·6464	0·6826	0·7245	0·7494	0·8183
2·00	0·6667	0·7434	0·8075	0·8323	0·8580	0·8721	0·9083
3·00	0·7500	0·8241	0·8750	0·8926	0·9100	0·9192	0·9423

As an example of particular interest the second line $z = 0\cdot461$ is the redshift for the most distant radio galaxy known 3C 295. If the universe is such that the deceleration parameter $q = +1\cdot0$, then the light from 3C 295 has taken 0·4730 of the time since the beginning of the expan-

* A. Sandage, *Astrophys. J.*, **134**, 916, 1961.

sion to reach us. That is, when observing this object we look back nearly half way to the beginning of the expansion. If the redshifts of the quasars are wholly associated with the expansion of the universe, then, when observing those of the largest known redshifts ($z = 2\cdot7$) we look back nearly 90 per cent of the way to the beginning of the expansion for a universe in which $q = +1$. The look-back times for the Einstein–de Sitter universe ($q = +\frac{1}{2}$) are significantly less, as can be seen from the table. No figures appropriate to the steady-state theory appear in the table since $q = -1$ and t_0 has no meaning.

(b) Luminosity–distance relations

The classical approach to the problem of distance determination of galaxies which lie beyond the range of detection of Cepheid variables, is to measure the apparent luminosity of the galaxy. If it is of an identifiable type for which the absolute luminosity is known, then, in principle, the distance can be determined. However, a rapidly receding object is fainter than it would be if stationary at the same distance (a burst of energy is received over a longer interval than the interval of emission because of the recession). Also, another factor of similar magnitude is introduced because, in the theory of relativity the energy emitted by a source is measured differently by an observer moving with that source, than by one situated elsewhere.

The relations between the apparent bolometric magnitude* (m_{bol}) of a galaxy, the deceleration parameter q and the redshift z have been given by Sandage†. For values of $\Lambda = 0$ and for $q > 0$

$$m_{bol} = 5 \log \frac{1}{q^2}\left\{qz+(q-1)[(1+2qz)^{\frac{1}{2}}-1]\right\}+C$$

For the case when

$$\Lambda = 0 \quad \text{and} \quad q = 0$$
$$m_{bol} = 5 \log z(1+\tfrac{1}{2}z)+C$$

For the steady-state model where $q = -1$,

$$m_{bol} = 5 \log z+5 \log (1+z)+C$$

The term C contains the Hubble constant and the total luminosity of the galaxy. It is constant only if the latter is the same for all galaxies in the sample under observation, and can be determined by fitting the equations to observations of bright galaxies. Then, if observations of m_{bol} to sufficiently high values of z could be made, the fundamental para-

---

* That is, the apparent radiation at all wavelengths. The *absolute* bolometric magnitude may be derived from the visual absolute magnitude, and the spectral type, and by making allowances for the atmospheric absorption.

† A. Sandage, *Astrophys. J.*, **133**, 355, 1961.

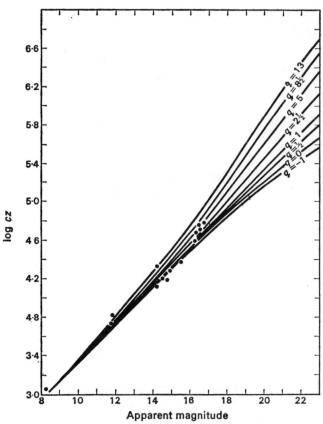

21. The theoretical relationship between the redshift z (the ordinate is the logarithm of zc with $c = 3 \times 10^5$ km s^{-1}) and the apparent magnitude for cosmological models with zero Λ and various values of the deceleration parameter q. The points refer to 18 distant clusters of galaxies for which the redshift/magnitude values were well established in 1961. The position in 1970 is shown in Fig. 22.

meters of the universe could be derived. The immense difficulty of the problem is well illustrated by Fig. 21, taken from the paper by Sandage.[*] This shows the relation between the redshift z (on a logarithmic scale) and the magnitude, calculated for models with zero cosmological constant and various values of the deceleration parameter q. The diagram shows that at a redshift of $z = 0.5$ c (the radio galaxy 3C 295 has a z value of 0.461) the predicted difference in magnitude between the steady-state universe ($q = -1$), and the Einstein–de Sitter universe ($q = +\frac{1}{2}$) is only 0.7 magnitude. Even so, such differences at these magnitudes should be detectable with the 200 inch telescope.

The points for well established measurements on 18 clusters are

* A. Sandage, *Astrophys J.*, **133**, 355, 1961.

shown on this diagram. It was hoped that the discovery of objects at values of z substantially greater than 0·5 would define the value of q without ambiguity—since at these greater values of z the various models depart significantly from the asymptotic line. This hope has not materialized. The quasars, with much greater redshifts extending to values of more than $z = 2\cdot5$, show an even greater scatter on this type of Hubble diagram. Thus, so far, no significant clarification has occurred into the value which should be adopted for the deceleration parameter. This problem is illustrated in Fig. 22 which shows the redshift/apparent

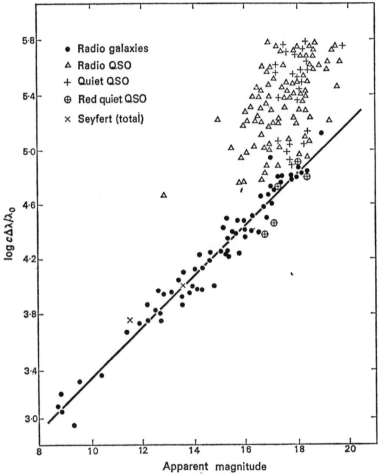

22. The observed redshift/apparent magnitude diagram for radio galaxies, quasars, and other quasi-stellar objects, as established to 1970. The line corresponds to a model with $q = +1$. (The ordinate is the logarithm of zc with $c = 3 \times 10^5$ km s^{-1}.) 'Quiet QSO' are the quasi-stellar objects without radio emission. 'Red quiet QSO' are quasars redder than the average quasar at the given redshift.

magnitude diagram for radio galaxies, quasars, and other quasi-stellar objects according to measurements available up to 1970. Although the radio galaxies lie near a line with a slope corresponding to $q = +1$, the quasars of high redshifts show a great scatter. However, this diagram illustrates a curious feature of the present situation. It is that there appears to be a sudden cut-off at redshifts of about $z = 2 \cdot 8$ (corresponding to log cz of about $5 \cdot 9$ in the diagram).* Of course, this could be some unknown observational selection effect but if eventually it should be found to be a real cut-off there are interesting consequences. If the deceleration parameter $q = 0$, then at $z = 3$, the look-back time into the past history of the universe is $0 \cdot 75\ t_0$ years to the singularity. In this case the implication of the cut-off would be that the quasars were born $0 \cdot 25\ t_0$ years after the singular condition and that when we look beyond this horizon we do not see more luminous bodies because they did not yet exist. For a deceleration parameter of $q = +1$, the look-back time at $z = 3$ is 89 per cent of the time to the singular condition, and the quasars would have been born $0 \cdot 09\ t_0$ years after this beginning.

(c) *Angular diameter–redshift relationship*

If we measure the angular diameter θ of an object with a well-defined linear diameter, then we expect that θ will decrease linearly as the distance r increases. For small values of r (small redshift z), this certainly applies to the measurement of the angular diameters, for example, of a class of galaxies having well-defined linear dimensions. However, at great distances (large z), the curvature of space–time associated with the k term in the field equations of general relativity introduces a new feature which changes the linear relationship between θ and z.

It can be shown that for an object with a given linear dimension the relation between the apparent angular diameter $d\theta$ and the redshift z is of the form

$$d\theta \propto (1+z)^2$$

the proportionality involving k and q. Thus if the variation of $d\theta$ with z can be established, the parameters of the universe could be found. It is precisely this argument which has stimulated much of the work on the angular diameter of the radio sources already described.

The above relation conceals a remarkable property of the universe. It can be expressed in terms of a parameter D proportional to the luminosity of the galaxy as

---

* There is also the cut-off along the abscissae at an apparent magnitude of about 20. This is, of course, a consequence of the limiting sensitivity of the telescopes. The point to be emphasized is the apparent cut-off at values of log $cz = 5 \cdot 8$ within the observable apparent magnitude range between 16 and 20.

$$d\theta = \frac{(1+z)^2}{D} \text{ for values of } q \geqslant 0$$

For small z $(z < 1)$ the relationship expresses a decrease of $d\theta$ as z increases. However, for large z, the prediction is that $d\theta$ will actually increase with increasing z. In fact, since D is nearly proportional to z, the theory predicts that in evolving universes the angular diameter will decrease as z increases, initially, will reach a minimum, and will then increase for large values of z.

On the other hand, the steady-state theory with $q = -1$, predicts a different variation of $d\theta$ with z of the form

$$d\theta \propto \frac{1+z}{z}$$

This does not go through a minimum as z increases, but decreases continuously with increasing z, tending asymptotically to a limiting value determined by the value of the Hubble constant and the linear dimensions of the source.

Reference has already been made in Chapter 4 (Fig. 5) to this striking difference between the predictions of the steady-state and evolutionary models which Hoyle discussed at the Paris Symposium on Radio Astronomy in 1958. The position of the minimum in the evolutionary models depends on the values of q and Λ. For instance, in the Einstein–de Sitter model with $\Lambda = 0$ and $q = \frac{1}{2}$ the minimum occurs at a z value of about 2. As an example of the angular sizes involved, Hoyle, in 1958, quoted the case of the radio galaxy in Cygnus. With a measured value of $d\theta$ of 80 seconds of arc at a redshift $d\lambda/\lambda$ of $\frac{1}{18}$, the minimum in the Einstein–de Sitter universe for objects of this size would be 15 seconds of arc, and the asymptotic value in the steady-state universe would be 4 seconds of arc.

These predictions of a striking difference between the various models of the universe stimulated hope that a decisive observational test would be possible with the radio telescopes. Although it was subsequently realized that the typical linear size was probably much smaller than the example quoted above, nevertheless there seems no insuperable technical problem in the execution of this observational test. The difficulty encountered up to the present time exists because no satisfactory means have yet been found of evading the problem presented by the wide range of linear dimensions of the radio galaxies and quasars. Efforts are in progress to attempt a delineation of a homogeneous class of objects from the many thousands which lie within the scope of angular diameter measurements with contemporary radio telescopes. When a sufficient number of objects of standard size can be isolated at various redshifts this crucial observational test seems bound to yield important information about the parameters of the universe.

(d) *Number counts*

At least as far as radio astronomers are concerned the most popular test of the cosmological theories has been based on the measurements of the number of radio sources as a function of distance. The basis of the test is that if galaxies, or other objects, are distributed uniformly in space at each cosmic epoch, then the field equations of general relativity can be used to predict the variation of numbers (N) per unit solid angle as a function of the redshift or luminosity. This relation between $\log N$ and the magnitude is shown in Fig. 23 (a) as calculated by Sandage* for the steady-state theory $q = -1$, and for evolutionary models with $q = \frac{1}{2}$ (Einstein–de Sitter), and $q = 2\frac{1}{2}$. In the radio counts the observations give the number counts N as a function of the flux density and these relations have been compared with the simplest illustrative case of a static universe containing n sources per unit volume with absolute radio luminosity P. If a radio telescope measures a number of sources whose luminosity exceeds S, then according to the inverse square law this number comprises all sources within a sphere of radius

$$\left(\frac{P}{S}\right)^{\frac{1}{2}}$$

The total number will be $\frac{4}{3}\pi n P^{\frac{3}{2}} S^{-\frac{3}{2}}$ and the number N per unit solid angle will be

$$N = \frac{1}{3} n P^{\frac{3}{2}} S^{-\frac{3}{2}}$$

A large spread in luminosity will not affect the three-halves power law so that if Σ is the summation over all luminosity classes

$$N = \frac{1}{3}(\Sigma n P^{\frac{3}{2}}) S^{-\frac{3}{2}}$$

Thus for a uniform distribution in a static universe $N \propto S^{-\frac{3}{2}}$ or $\log N/\log S = -\frac{3}{2}$. This $\log N/\log S$ relation is shown in Fig. 23(b) as a straight line of slope $-1\cdot5$.

The comparison of the number counts of radio objects with flux densities initiated by Ryle and his colleagues in Cambridge has consistently shown a curve steeper than the prediction of the three halves power law as shown in Fig. 23(b). Now all models of the type illustrated in Fig. 23(a), which are free of the simplifying assumption of the static universe, predict a $\log N/\log S$ curve which has everywhere a slope less steep than $\frac{3}{2}$. Arguments may be advanced to show that effects of the redshift as S decreases would tend to flatten the observed curve. Much discussion has, in turn, centred on the possibility of evolutionary effects

* A. Sandage, *Astrophys. J.*, **133**, 355, 1961.

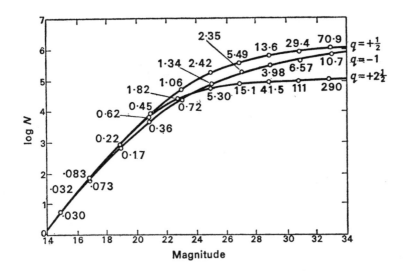

23(a) The theoretical relations between the number counts N (plotted as log N) and the magnitude for the steady-state model ($q = -1$) and for two evolutionary models $q = \frac{1}{2}$ (the Einstein–de Sitter model) and $q = +2\cdot5$. The numbers on the curves are the equivalent values of the redshift z.

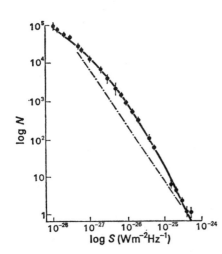

23(b) The solid line shows the Cambridge radio source counts on 408 MHz published by C. G. Pooley and M. Ryle in 1968. The broken line has a slope of $-1\cdot5$. The ordinate is the logarithm of the number (N) of sources per unit solid angle whose flux density (S) exceeds the values plotted as log S on the abscissa.

on the observed curve—that is, the objects with low values of S, being the most distant, are observed during an early evolutionary state of the universe, and that in the past the sources had a greater absolute luminosity. By this form of argument it is possible to bring the theoretical and observed number count curves more into alignment. Of course, such an explanation of the measurements would be quite incompatible with the steady-state theory. The steady-state theory makes a unique prediction of a log N/log S curve with a slope less steep than $\frac{3}{2}$, and since the theory

inhibits the introduction of an evolutionary-distance effect the observations have been held to be decisively in favour of an evolutionary
model.

The successive measurements and arguments presented by Ryle and
his colleagues have been in favour of an evolutionary universe although
it must be mentioned that the topic of number counts has led to much
dispute in recent years. Several other independent measurements of the
log N/log S relationship have not shown precise agreement with the
Cambridge results. It is held, for example, that not all possibilities of the
effects of inhomogeneities in the samples have been excluded. In fact,
the detailed uses of the technique for the determination of the parameters of the universe are hindered by doubts about the observational
selection effects and by a lack of understanding of the nature of possible
evolutionary effects in the distant regions of the universe.*

(e) *The microwave background radiation*

Although no measurements made at Jodrell Bank have been concerned
with the problem of the microwave background radiation it seems
desirable to refer to this topic since many astronomers regard the discovery as a decisive test of cosmological theories. During 1964 and 1965
scientists of the Bell Telephone Laboratories, New Jersey, were concerned with the development of receiving equipment of high sensitivity
in the centimetre waveband for communication tests using the Echo
balloon satellite. In order to avoid effects of ground emission a special
horn type aerial system had been developed so that signals received
outside the main beam, in side-lobes, were much weaker than with the
more conventional paraboloid. In spite of these precautions it was
found that the noise level in the equipment when directed at the sky was
100 times greater than it should be from known sources of galactic or
extragalactic radio emission at those wavelengths. Further, A. A.
Penzias and R. W. Wilson, who were primarily concerned with these
measurements, discovered that this excess noise was isotropic over the
sky to a few per cent. They announced that at this wavelength of 7 cm,
there existed an isotropic background radiation equivalent to a black
body temperature of 3·5 K (later modified to 3·1 K). Several other
workers soon confirmed this result at neighbouring wavelengths and
the effect became known as the '3 degree background radiation'.

* Criticisms of the cosmological interpretation of the source counts have been made
by F. Hoyle in his 1968 Bakerian lecture (*Proc. R. Soc. London*, Ser. A., **308**, 1, 1968)
and by K. I. Kellermann in his 1971 Helen B. Warner lecture to the American
Astronomical Society (*Astr. J.*, **77**, 531, 1972). An account of the various arguments
in favour of the cosmological interpretation in terms of an evolutionary universe has
been given by M. Ryle (*A. Rev. Astr. Astrophys.*, **6**, 249, 1968).

1. The author (*right*) and Professor J. G. Davies (*left*) explaining to a press conference at Jodrell Bank on 18 October 1967 the results of the observations of the Soviet probe to Venus.

2a. The Mk II radio telescope at Jodrell Bank. The smaller telescope to the left is the 50 ft polar axis mounted instrument and the one to the right on top of the power house is the small 25 ft radio telescope originally built on top of the Shot Tower on the South Bank for the 1951 Exhibition.

2b. The Mk III radio telescope at Wardle near Nantwich, Cheshire, controlled by radio link from Jodrell Bank. The bowl of this telescope is the same size and shape as that of the Mk II (*above*), but the surface is of mesh. In this photograph the bowl is facing the zenith.

3. The Lick Observatory stroboscopic television photographs taken on 3 February 1969 of the light pulses from the Crab nebula pulsar PSR 0531+21. The photographs show the region near the centre of the nebula: top left when the pulsar is at maximum brightness; bottom left at minimum. The two photographs on the right show the pulsar at intermediate brightness. The total exposure was ten seconds for each photograph.

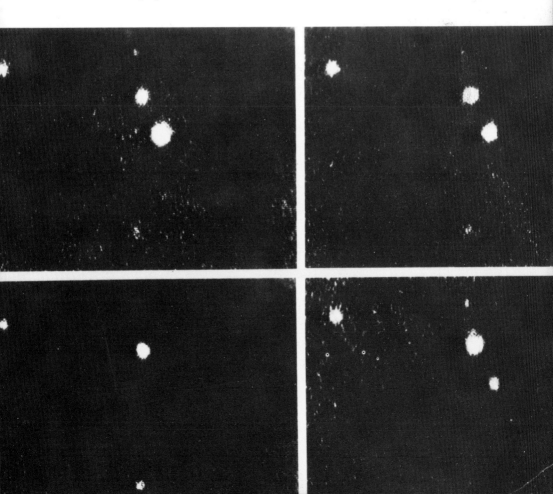

4a. Dr. J. E. B. Ponsonby (*right*) and the 'ephemeris doppler machine' which he devised for the Venus planetary radar observations. His assistants are M. Gallagher (*left*) and R. Porter (*centre*).

4b. Sir Martin Ryle (*left*) and Professor Graham Smith photographed in the early 1960s at Lord's Bridge, Cambridge, near the aerials of the radio telescope with which the 3C survey of radio sources was made. Professor Graham Smith came to the University of Manchester and Jodrell Bank in 1964.

5. The planetary system as drawn by J.G. Davies (*above*) and as received (*below*) at Zimenki Observatory in the U.S.S.R. after transmission via the Jodrell Bank telescope and the Echo II balloon in February 1964.

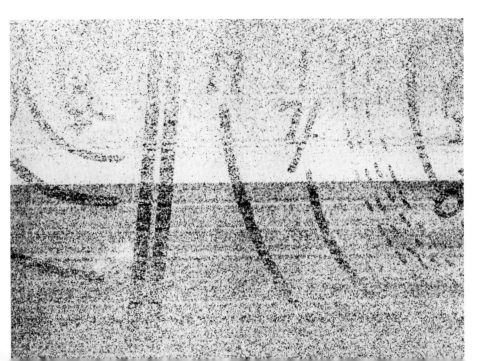

6. The receiving laboratory at Zimenki Observatory in the U.S.S.R. during the Jodrell Bank–Echo II–U.S.S.R. communication experiments in February/March 1964. The Soviet scientists in this photograph are (*left to right*) Engineer R. Purckhovsky, leading engineer of computing centre V. Kalinin, leader of radio telescope operation group I. Puzyrev.

7. The MK I radio telescope (*above*) and the Mk IA (*below*). The major changes in the conversion to the Mk IA were the replacement of the original stabilizing girder by the heavy load-bearing wheels and the construction of the new reflecting surface above the original membrane which is still retained.

8. The Mk I telescope during the conversion to the Mk IA, photographed on 20 April 1971. The new load-bearing wheel was nearing completion. The panels of the new membrane are stacked on the corner of the grass field near the base of the tower crane. The holes in the original membrane were cut to allow the supporting framework for the new reflector to be fixed to the main elevation girders.

The control building is at bottom left adjacent to the 50 ft telescope. The building to the bottom right is the public concourse building and planetarium which had been officially opened by the chancellor of the University (His Grace The Duke of Devonshire) the previous day.

A remarkable feature of this discovery was that Professor R. H. Dicke and his colleagues at Princeton, New Jersey, had been simultaneously considering the implications of a high temperature collapsed phase of the universe—either as a singular event or as the epoch of maximum collapse in an oscillating universe. From considerations of the present density they had concluded that the temperature at such an epoch would be 10^{10} K, and that it should be possible now to detect the relic black body radiation which they calculated would have a temperature less than 40 K. They had built apparatus working on a wavelength of 3 cm to attempt this measurement, but before results were obtained they learnt of the successful measurements of Penzias and Wilson at Holmdel, New Jersey.*

The theory of the possible high temperature phase of the universe, its subsequent thermal history and the mode of production of the elements from the primeval matter then published by Dicke was not the first attempt to explain the present state of the universe on the basis of an initial high temperature condition. The first idea of this type was proposed by Gamov in 1946 and was subsequently developed by him in collaboration with Alpher and Bethe.† Gamov accepted that the universe began in a singular condition of high density according to the point source evolutionary models already discussed. His special contribution was to propose that the early dense stages in the initial moments of the expansion were hot enough for the formation of the elements to occur through thermonuclear reactions. Under the original enormous pressure the universe was assumed to consist primarily of neutrons, and following the explosive beginning when this great ball of neutrons began to expand, the decay of neutrons into protons and electrons would occur. The half-life for decay of a neutron is 700 seconds so that 700 seconds after the explosion one half of the original neutrons would have decayed into protons and electrons if no further nuclear reactions occurred. However, the temperature near the beginning on this theory would have been exceedingly high—after 1 second, about 10 000 million degrees. After the 700 seconds in which half the neutrons would have decayed, the temperature was still about 500 million degrees. Thus, in these early stages of the expansion thermonuclear reactions could take place between the remaining neutrons and the protons and electrons of the decay products and, in the first minutes of the life of the universe, the

* These results were published simultaneously by R. H. Dicke, P. J. E. Peebles, P. G. Roll, and D. T. Wilkinson, *Astrophys. J.*, **142**, 414, 1965, and A. A. Penzias and R. W. Wilson, *Astrophys. J.*, **142**, 419, 1965.

† Since the key paper was published by Alpher, Bethe and Gamov, the theory became known as the α-β-γ theory.

elements would be built up by successive capture of neutrons. On this theory the entire process of element building must have occurred in about the first 30 minutes of the life of the universe. Then the temperature would have fallen below the threshold at which the thermonuclear reactions can occur.

Although the formation of the elements is not our present concern, it must be mentioned that the concept that all the elements could be produced in the first minutes of the life of the exploding universe is no longer regarded as tenable. An important difficulty is that in the sequence of the build-up of elements by neutron capture there is no stable atom of mass 5 or mass 8. If helium 4 is bombarded with neutrons, the helium 5 produced immediately breaks down to helium 4 again. Similarly, if the beryllium isotope of mass 8 is produced it instantly breaks down into two helium 4 atoms. Thus, it seems that in the theory of the production of the elements by neutron capture in the first half-hour of the life of the universe the build-up could get no further than helium.

The Gamov hot big bang theory provides a dramatic picture of the first moments of the universe, but the fundamental fallacy appears to be the assumption that neutrons and their natural decay were of predominant importance in the early stages. In 1950 the Japanese astrophysicist, Hayashi, pointed out that in any such singular condensed phase the temperature would be greater than 10^{10} K which exceeds the threshold necessary for the creation of electron–positron pairs. In this event the neutron transformations would not be governed by their natural half-life of 700 seconds, but by processes involving the interaction of high energy positrons. The neutron abundance would be materially altered within a second at these temperatures and almost instantly thermal equilibrium would be established between matter and radiation. The consequences of this early decoupling of matter and radiation, and of the survival of photons from the early stage of the universe, has been considered in much detail by R. C. Herman, W. A. Fowler, F. Hoyle, and others since the original proposals of the α-β-γ theory.

The currently accepted view of the production of the elements is that they are formed by thermonuclear reactions in the hot interior of the stars. Nevertheless there is a problem about the abundance of helium in the universe. In our own galaxy, at least, 92 per cent of the atoms are hydrogen, 8 per cent helium, and only about one atom in a thousand represents other heavier elements. The relative abundance of helium leads some astronomers to the view that the element building must have been a combination of element formation to the helium stage in the high temperature state of the universe and then the later formation of the heavier elements in the stellar interiors.

Our main concern here, though, is with the predicted background temperature on the concept that the universe either began with a 'hot big bang' of this nature or has evolved from a collapsed phase of high temperature and density. The rate of cooling of the universe would have been determined by the relative influences of radiant energy and matter. On the Gamov model the radiant energy decreased rapidly as the universe expanded (proportional to the fourth power of the scale factor R). He calculated that after 250 million years radiation ceased to be the dominant factor in the universe; at that time the density of ordinary matter became equal to the mass density of radiant energy. Then great proto-galaxies of gas clouds emerged and subsequently condensed into galaxies and stars. As for the change in temperature we have already mentioned the figure of 10 000 million degrees after 1 second, and 500 million degrees after 700 seconds. The continuous expansion and the relatively changing roles of radiant energy and matter would lead to a decrease to about 6000 degrees after 200 000 years, and to only about 170 degrees above the absolute zero at this epoch of change-over from the radiation-dominated to the matter-dominated universe. The more precise calculation of these temperature changes with the age of the universe (and of the present-day value) depend critically on the relative amount of radiation and matter during the evolutionary process, and as already mentioned, the assumptions made in the α-β-γ theory are no longer believed to be justified.

The original calculations made by Gamov and his collaborators predicted a present-day black body temperature of 25 degrees absolute. Twenty years later Penzias and Wilson assumed that they had observed this isotropic microwave background emission equivalent to a black body of about 3 degrees absolute. Although by nearly a factor of ten lower than Gamov's original prediction, it was realized that the observation and prediction were not necessarily incompatible, because of the sensitiveness of the prediction in this and the subsequent theories to the assumptions about the early helium production. It is understandable, therefore, that several other astronomers quickly checked this discovery and made measurements on other wavelengths in an attempt to establish that it did indeed have the spectrum appropriate to a black body at that temperature. The present position is that measurements have been made covering the wavelength range downwards from about 20 cm to about 3 mm. The relation between wavelength and the measured background intensity corresponds closely to the line appropriate to a black body temperature of approximately 2·7 K (that is somewhat lower than the original figure given by Penzias and Wilson). Black body radiation at this temperature has a peak intensity in the 1 mm wavelength region and a most critical test would be to establish that the observations showed a peak at this wavelength. There are, however, profound difficulties in making measurements at these millimetre wavelengths—not only because of absorption effects in the Earth's atmosphere, but also because many interstellar molecules have line spectra in this wavelength range. It may be necessary to make observations from space vehicles in order to evade the problems introduced by the Earth's atmosphere. In that case if the measurements showed a maximum in the vicinity of a millimetre wavelength it would seem, at this moment, difficult to escape the conclusion that the radiation was, indeed, a relic of the original 'primeval fireball' or of a high temperature collapsed phase of the universe. Until that can be done, and until the complex arguments about the reasons for the precise isotropy can be settled, many astronomers are reserving judgement on the ultimate interpretation of this remarkable discovery.*

* Early in 1973 a group of scientists from the Los Alamos laboratory of the University of California published the results of measurements made by an infrared radiometer cooled by superfluid helium, launched by a rocket from the Kauai Test Range, Hawaii. The bolometer extended the measurements of the background radiation down to 0·4 mm, and the results were consistent with a 2·7 K black body background (K. D. Williamson, A. G. Blair, L. L. Catlin, R. D. Hiebert, E. G. Loyd, and H. V. Romero, *Nature, Phys. Sci.*, **241**, 79, 1973).

10

The discovery of pulsars

So far in this book I have described a connected sequence of researches, which although significantly influenced in their course by unexpected events, nevertheless represented a chosen line of work. Early in 1968 our calm pursuit of these researches with the telescopes was disturbed by a remarkable piece of news. In the scientific world, particularly in astronomy, the national and international circulation of individuals and news is so free that there are scarcely ever startling surprises. Even in the case of the Soviet Sputnik of October 1957, although the world as a whole was amazed, those scientists who were interested in such possibilities could have had no doubt at least of the Soviet intentions. The discovery of pulsars came to me and my colleagues as an abrupt and almost incredible surprise as the following account of my own initiation illustrates.

The Secretary of State* was already seated on the right of the Chairman of the Science Research Council. The noise of the remaining members settling around the Council table was subsiding as Fred Hoyle slipped into the chair on my right. He had returned only a few days earlier from America. 'Any news?' I whispered, meaning was there any fresh news about quasar identification. 'Not much', he replied, 'still more odd things about the absorption lines.' Then, almost as an afterthought, he added, 'but last night at a colloquium in Cambridge, Hewish said that he had discovered some radio sources which emitted in pulses with intervals of about a second'. 'Impossible,' I said; 'he must be picking up some odd form of interference.' 'No,' Fred added; 'the evidence for extraterrestrial origin seems convincing.'

The whispered conversation had to cease at that tantalizing moment.

* At that time the Secretary of State for Education and Science was the Right Hon. Patrick Gordon Walker, P.C., C.H., M.P. He had succeeded the Right Hon. Anthony Crossland, P.C., M.P. in August 1967 and was himself succeeded by the Right Hon. Edward Short, P.C., M.P., in April 1968.

The Chairman* was beginning to speak, explaining to the Secretary of State that this was a meeting at which the Science Research Council would hear a review of the work and proposals of the Astronomy Space and Radio Board, and that as the Chairman of the Board I would introduce this review. So I had to concentrate on my speech and try to forget this amazing piece of news. The occasion was the meeting of the Council on 21 February 1968 and we had produced a fairly substantial document in which we had covered the whole area of our research interests in space as well as in optical and radio astronomy.† We were also asked to outline our proposals for future developments and it was at this meeting that I explained that because of the economic restraints imposed on my Board we could not possibly proceed simultaneously with the Jodrell Mk V telescope and the Cambridge 5 km aperture synthesis instrument. Because of this we had decided to go ahead as quickly as possible with the Cambridge telescope, and delay the Mk V for two years so that the peaks in the expenditure would not coincide. At the same time we proposed to give immediate priority to the repairs and improvements to the Mk I at Jodrell Bank so that its useful life would be extended until the delayed Mk V became available.‡

As he passed me on the way out of the Council Room, the Secretary of State said, 'That was a statesmanlike decision'. When he had passed beyond hearing Fred Hoyle said, 'You heard that—you'll get the Mk V as well as the Mk IA'. I was not so sure, but in any case at that moment I was far more fundamentally disturbed by our previous conversation about the pulsating radio emissions and immediately returned to the subject. Fred Hoyle then told me that Hewish had been investigating these pulsed emissions for months, that four had been discovered and that their motion in the sky and other characteristics pointed conclusively to an origin in the Milky Way and that an account would appear in *Nature* on Saturday.

On the journey home I could think of little but this fantastic news. Some of my people at Jodrell must surely have known about this. Why hadn't I been told and why weren't we checking these objects with the telescope? In fact, it turned out that they had no information at all. The truth is that Hewish and the whole Cambridge group had for several months achieved a screen of security and secrecy which, in itself, was almost as much of an accomplishment as the discovery itself. Not that I

* Professor Sir Brian Flowers, F.R.S., who succeeded Sir Harry Melville in this post in 1967.

† The main part of this document was subsequently published as *Astronomy Space and Radio Board's Review* by the Science Research Council in July 1968.

‡ See Ch. 16.

blamed them for this. On the contrary, the reaction of the Press to the news—imputing an intelligent origin in the galaxy for the signals—showed only too clearly what could have happened if the news of the strange signals had leaked out before Hewish was able to produce the convincing scientific arguments for the natural origin of the signals.

The story of the discovery of the pulsating radio sources is, like the story of the discovery of the radio emissions from space and of many other events in science, one of an accidental and unintentional observation. The research programme which brought evidence of their existence to human eyes was not even the main research interest of Martin Ryle and his group. The immediate line of development began in 1964 when Hewish and others discovered the phenomenon of interplanetary scintillation.

If a radio telescope receiving signals in the metre wave band is used to record the radio emissions from a distant radio galaxy or quasar the record obtained is normally steady. It is, of course, true that in the early postwar days when the subject of radio astronomy was beginning to develop, Hey and his colleagues found that in the direction of Cygnus the radio waves were fluctuating in intensity with periods of several seconds or minutes. In fact Hey deduced from this that in the direction of Cygnus there must be a small-diameter radio source which was itself varying in strength. The conclusion was correct although the reasoning was wrong. It was soon shown by observing the source simultaneously by means of spaced receivers that if the receivers were close together then the fluctuations occurred simultaneously on both. On the other hand, if the receivers were separated by a large distance (between Cambridge and Jodrell Bank) then there was no correlation. These experiments proved that the variable signals arose because the steady emission from a source of small angular diameter in Cygnus was diffracted by the irregularities of the ionization in the F region of the Earth's ionosphere. As soon as the mechanism was understood, the technique became an important method for measuring the rate and direction of the drifts in the ionosphere. The effect was analogous to the twinkling of an ordinary star. Here the diffraction of the light from the star occurs because of irregularities in the density of the lower troposphere. If the source has a sufficiently large diameter, such as a planet, then twinkling does not occur. Similarly with the diffraction of the radio waves in the ionosphere, only sources of sufficiently small angular diameter show the effect—the radio emission from a planet does not.

The interplanetary scintillation discovered by Hewish in 1964 was another similar effect. In this case again only radio sources of small angular diameter scintillated, and the variations were of much shorter period than those produced by the ionospheric diffracting screen. The timescale of the fluctuations was only of the order of seconds, and it was soon shown that these fluctuations arose when the radio waves from the source traversed the plasma clouds of ionized particles in interplanetary space. It was soon realized that this discovery provided an important new method for setting upper limits to the angular diameters of many of the radio galaxies and quasars which had not yet been resolved by the long baseline interferometers. Furthermore, in its annual journey across the heavens the Sun and its extended corona crossed the line of sight of many radio sources. In the case of those sources which were small enough to show the scintillation effect and whose diameters were known, a study of their scintillating emissions as they moved through the solar corona was soon yielding new information about the structure of the corona out to many solar radii.

When a new effect is discovered the experimenter is often quick to realize that much more could be achieved if money were available to build better apparatus. So it was with Hewish. It was not long before the D.S.I.R.* received from him an application for £17 286 to enable him to build a large aerial system specifically to exploit this new discovery. He had produced convincing arguments that large numbers of additional radio sources could be studied and much would be learnt about angular diameters of the radio galaxies and quasars and also about the structure of the interplanetary medium, if he could construct a large aerial system. The plasma effects involved were more pronounced the longer the wavelength. He therefore chose to work on 3·7 metres (81·5 MHz). To attain the necessary sensitivity this meant a large collecting area and he proposed to build a fixed rectangular aerial array of 2048 dipoles spread over 4½ acres. An aerial of this size fixed to the ground would cover only a narrow strip of the sky, and the cost of making such a structure steerable would have been prohibitive. Hewish therefore proposed to use a pre-set arrangement whereby the beam could be steered in declination by changing the phases of the aerials. The rotation of the Earth scanned the aerial beam in right ascension. The phasing arrangements were such that, in declination, beams of half width ±3° were produced at 3° intervals and four receivers were used so that four different declinations were observed simultaneously. In right ascension the beam width was ±½°, so that as the rotation of the Earth swept the beam over the sky each source would remain in the aerial beam for about 3 to 4 minutes.

---

* The Department of Scientific and Industrial Research was disbanded early in 1965. The part of its activities concerned with research grants to Universities was taken over by the newly formed Science Research Council.

I was Chairman of the D.S.I.R.'s Astronomy sub-committee when Hewish made his application*. The application came before us on 10 March 1965 as one of those from Martin Ryle's group. The three totalled £37 074. We had little money left in our minor grants allocation and it was essential for us to give priority to one of the other requests since the continued support of members of Ryle's group was involved. It was my personal view that Hewish's proposal was excellent and should be financed—I do not recall that anyone dissented. On the other hand we did not have the money and had no alternative but to hold over the application. On 12 March 1965 I wrote to Martin Ryle to explain our dilemma since I knew from bitter experience how chilling and disconcerting the official negatives can be. 'Our next formal meeting is not until the summer, but after next week's Research Grants Committee session we are hoping that there will be a sufficient increase in our allocation for me to take action on Hewish's application. It might therefore happen that the delay in effect is only a few weeks.' I went on to say that my only anxiety was about his chosen wavelength of 3·7 metres, since we now found that to be unusable at Jodrell because it had been allocated to public service vehicles. Martin Ryle replied to me on 16 March and gave me the necessary assurance that this waveband was still clear at Lord's Bridge. My hope that the delay would be shortlived was justified and on 23 March I wrote again: 'I am very glad to be able to tell you that all obstacles have been removed.'

In the event the construction of the aerial and equipment was completed in 1967 and routine recordings began in July of that year. The short period of the scintillation to be studied led to the design of the receiving equipment such that the time constant was only 0·1 seconds. This unusually short time constant (of the order of ten times shorter than that normally employed in sky surveys) coupled with the high sensitivity of the system—the noise fluctuations corresponded to a flux density of only $0·5 \times 10^{-26} \text{W m}^{-2} \text{Hz}^{-1}$—produced quite fortuitously the ideal conditions for revealing the existence of the pulsars. The declination phasing arrangements made it possible to scan the whole sky in the range of declination $-08° < \delta < 44°$ within one week and it was on this repetitive basis that the recordings began in July 1967.

* On the formation of the Science Research Council early in 1965 I became Chairman of the Astronomy Space and Radio Board which through the Astronomy Policy and Grants Committee assumed responsibility for the astronomy grants previously handled by the D.S.I.R.'s Astronomy sub-committee. However, the transfer took several months and Hewish's application was in fact, dealt with by the D.S.I.R. sub-committee. It was the last meeting but one—the final meeting was on 19 May 1965.

The record of the signals received by this equipment was in the form of a pen recording and the paper charts accumulated at the rate of about 400 feet per week. One of Hewish's research students, Miss Jocelyn Bell,* had the task of making the part of the analysis of these records concerned with the celestial co-ordinates of any scintillating component of the received signal. It is fortunate that Miss Bell was a faithful analyst and not one influenced by preconceived ideas of results to be expected, because on these records there appeared a scintillating source in transit near midnight, which is a time when the interplanetary scintillation effect becomes very small. By the end of August she had plotted this source on several occasions. Although the source was conspicuous on the records it was sporadic. This caused it to appear at random times in the few minutes of each night during which the source was in the aerial beam, giving the impression that the right ascension varied irregularly. For some time Hewish assumed that some form of terrestrial interference was present. By the end of September, however, it became clear that this irregularity in right ascension was only apparent and was an effect arising because of the random nature of the signals. Furthermore, there was no parallax of the magnitude to be expected if the signals were an interfering effect from a deep space probe, and by that time Hewish had concluded that the radio source responsible for this scintillating effect on the records must lie outside the solar system.

Although at that stage the evidence for an extraterrestrial origin of the signals was conclusive, their nature was still not realized. There are various radio sources in the galaxy which emit sporadically—such as flare stars—and further investigations were made to discover more about the nature of these signals. During October and November the signals were weak, but on 28 November satisfactory recordings of strong signals were obtained. These revealed for the first time that the signal was not emitted continuously by the source but in the form of pulses. High-speed recordings indicated that the source was emitting radio energy in the form of pulses, each pulse being about 300 milliseconds long and recurring every 1·337 seconds with a precision better than one part in ten million.

One of the important observations made on the pulsed emissions at this stage was the recording of the pulses simultaneously on two identical

* Subsequently Mrs. Jocelyn Bell-Burnell. When I heard that Miss Bell was the student who had first plotted the pulsating source I said to H. P. Palmer who was the member of my staff responsible for dealing with postgraduate admissions to Jodrell Bank, 'Surely Miss Bell wanted to work at Jodrell. Why is she at Cambridge?' to which Palmer replied, 'Don't you remember. Unfortunately I lost the correspondence so she decided to go to Cambridge instead.' To my irritated rejoinder he merely said: 'Anyhow if I hadn't lost the correspondence pulsars wouldn't have been discovered!'

receivers tuned to frequencies of 80·5 and 81·5 MHz. The pulses appearing on the lower frequency were delayed by 0·2 s compared with those received on 81·5 MHz. This corresponds to a frequency drift of -5 MHz s^{-1}. The most likely explanation which soon received confirmation from measurements elsewhere over much larger frequency intervals, was that this effect arose because of dispersion in the interstellar medium. The impulsive nature of the received signal is caused by the periodic passage of a signal of descending frequency through the narrow passband of the receiver.

From these dispersion measurements, and by assuming a value for the electron density in interstellar space, it was possible to make an estimate of the distance of the pulsating source. Even this preliminary work showed that there could be little doubt that the pulsating source must lie somewhere amongst the stars of the galaxy. Furthermore, the small instantaneous bandwidth of the signal and the rate of sweep showed that the duration of the emission at any given frequency could not exceed 0·016 s, hence it was concluded that the emitting source must have exceedingly small physical dimensions—not exceeding $4·8 \times 10^3$ km. Thus the source was only of planetary dimensions, but at stellar distances and emitting energy in a single pulse of at least 10^{17} erg if the radiation was isotropic. Hewish and his colleagues suggested straight away that the most likely candidates were either pulsating white dwarfs or neutron stars.

Naturally, the records which had yielded this remarkable information were searched for other pulsating sources as soon as the galactic nature of the first was established. Miss Bell re-examined the three miles of charts and within a few weeks three other pulsating sources were found in widely differing parts of the sky. At this stage, when Hewish and his colleagues were certain that the pulsating signals arose from some kind of stellar objects in the galaxy, they published the information—the first communication by himself, Miss Bell, J. D. H. Pilkington, P. F. Scott, and R. A. Collins was received in the offices of *Nature* on 9 February 1968.

An extraterrestrial civilization?

When this communication appeared in *Nature* in mid-February astronomers everywhere were amazed. Continuity was the order of the day in astronomical observations. True, sporadic outbursts in the optical and radio wavebands were well known—supernovae and solar flares for example—but these were still emissions of continuous radiation. The idea that celestial sources could produce pulses of energy required a fundamental reorientation of thought. It is perhaps scarcely surprising that many individuals, including a number of scientists, thought in terms of an extraterrestrial civilization as the origin of these pulses.

The amazement at the publication of the results was accompanied by a good deal of annoyance that the Cambridge scientists had behaved with such secrecy about their discovery. Many American scientists were especially vociferous in their disapproval, particularly when it became known that the Cambridge team had discovered three other pulsating sources but refused to publish the celestial co-ordinates of these—in fact the communication containing this information was not received by *Nature* until 3 April.

About two or three weeks after the publication of the discovery, Graham Smith who was Secretary of the Royal Astronomical Society, told me that feelings were strong on this issue and asked me what I thought of the situation. I replied that I thought the Cambridge people had behaved with exemplary scientific discipline in withholding news of their discovery until they were satisfied about the general nature of the objects, but that there was no excuse for a similar delay in withholding information about other similar objects which they had discovered.

It must be remembered that Hewish and his colleagues were faced with a quite extraordinary situation. When the signals were first noticed in the summer of 1967, the natural reaction was that they were some form of terrestrial interference which is the plague of every radio astronomer. Slowly it became clear that the origin of the signals was somewhere in the sky outside the Earth's environment. It was the end of November before the pulsating nature of the signals was established and it must have been hard even then to believe that they had a natural origin. Furthermore, at that time other evidence showed that the pulses came from an object whose dimensions were about the same as those of a planet, and moreover that the pulses were not constant in strength but variable. It was then necessary to make observations to find out if, indeed, this variability was a code, and whether the signals had an intelligent extraterrestrial origin.

Little imagination is required to conceive of the tumult which would have arisen if the information had been released at that stage. A gullible world would have seized on the extraterrestrial civilization concept and a scientific discovery of enormous fundamental importance would have been submerged in a torrent of science fictional nonsense. It is also quite certain that the scientific work of the Cambridge group would have been stopped for a long period by the inundation which seems to be associated with 'newsworthy' scientific items in our age.

It must have been very hard indeed for Hewish and his students to keep to themselves this unfolding drama. That they did so until they were able to produce the incontrovertible scientific evidence that the objects were amongst the stars, that there was no coding, and that the energies involved were those to be associated with natural phenomena, is to their very great credit and in the highest tradition of scientific research.

II

Jodrell Bank and the pulsars

At the time of my conversation with Fred Hoyle at the meeting of the Science Research Council, when I first heard about the Cambridge discovery, the Jodrell Bank telescope was working on three frequencies simultaneously—151, 240, and 408 MHz. It was being used for two research programmes: a night programme of my own was concerned with the radio emissions from the flare star YZ CMi (Ross 882)* and during the daytime it was in use by B. J. Rickett, who was studying the interplanetary scintillation phenomenon, specifically as a means for investigating the condition of the extended corona of the Sun. Normal pen recording charts were employed but in addition an Argus 400 digital computer was in use for the on-line analysis of the three receiver outputs. Another student, A. G. Lyne was closely associated with this equipment, although his main interest was in the study of the lunar occultation of radio sources.

I arrived home late on the Wednesday night after the meeting in London and came to Jodrell on the Thursday morning assuming that my colleagues would by this time have heard about the pulsating sources, and that they would be urgently engaged in turning the telescope onto the position of the source revealed by Hewish. The telescope seemed to me to be ideally equipped to obtain much additional information about the pulsating source, particularly to extend the wavelength range (since Cambridge had only observed on 80·5 to 81·5 MHz) and the dispersion measurements. I said that in my view the potential value of this information was far greater than that of my flare star observations and I offered to forgo my claim to the rest of the programme scheduled for the flare star work. My colleagues were strangely unenthusiastic. Rickett wanted to continue with his scintillation observations and both he and Lyne were reluctant to be distracted from the task of completing the observations necessary for them to prepare their theses for the Ph.D. degree. However, on Friday morning, Graham Smith told me that he had telephoned Hewish, who had given him the co-ordinates and the

* See Ch. 13.

period of the pulsar. He had 'taken fire' on the pulsar idea, and after I had reassured Rickett and Lyne that any delay to their Ph.D. completion would in no way be held against them—the telescope was set to work on this new problem. Before the issue of *Nature* containing the announcement of the Cambridge discovery was delivered at Jodrell Bank on the Saturday morning, the pulsating source was being received and studied on these higher frequencies. The amount of information which then flooded in, as soon as these advanced analytical techniques were brought to bear on the problem, raised great enthusiasm, and a week later two or three other individuals spontaneously abandoned their immediate tasks to help with the analysis and also to look for the signals with the Mk II telescope in the 11 cm waveband. Whereas the Cambridge telescope could only observe the pulsating source for about 4 minutes per day, the Jodrell telescope could follow it continuously as long as it was above the horizon and consequently information about the pulse shape and fading characteristics accumulated rapidly. After only two weeks work these new data were published*—the second of the great flood of communications which were soon to appear in *Nature* from many scientists all over the world.

Our own ability to contribute significantly to this work depended on two factors. Firstly on the steerable telescope with its high gain on short wavelengths, and secondly on the digital computer and data handling system which we had recently brought into use as described in Chapter 8. This had a capacity for data acquisition of an order of magnitude greater than that required for the existing astronomical observations and flexible programming arrangements which facilitated the immediate adaptation of the telescope and computer system to this new problem.

The delay observed by the Cambridge team in the arrival of the pulses on two frequencies separated by 1 MHz (81·5 to 80·5 MHz), had only two reasonable explanations, either the delay was occurring in the source itself—as in the phenomenon of the frequency drift bursts emitted by the Sun when it flares—or the frequency drift was caused by dispersion in the interstellar medium. In their initial communication of the results Hewish and his colleagues thought that the interstellar dispersion effect was the more probable. All the data suggested that the pulses were originating in a highly condensed object such as a white dwarf or neutron star. In this case the scale height of any atmosphere would be small and a travelling disturbance would be expected to produce a much faster frequency drift than that observed of 4·9 MHz s^{-1}

* J. G. Davies, P. W. Horton, A. G. Lyne, B. J. Rickett, and F. G. Smith, 'Pulsating radio source at $a = 19^h 19^m$, $\delta = +22°$', *Nature*, **217**, 910, 1968 (received in the *Nature* office 2 March, publication date 9 March 1968).

which is of the same order as that found for the frequency drifts in the solar atmosphere.

The Jodrell observations extended the drift measurements to 408 MHz and made the first of these alternatives even less likely. If a radio wave of frequency ν passes through an ionized medium (such as that of interstellar space), then there will be a frequency drift $d\nu/dt$ given by

$$\frac{d\nu}{dt} = -\frac{c}{L}\frac{\nu^3}{\nu_p^2}$$

When L is the path length, c the velocity of light and ν_p the plasma frequency $\left(\nu_p^2 = \dfrac{Ne^2}{\pi m}\right)$. Thus, if a source in space emits over a wide frequency band simultaneously, then because of this dispersion, the higher frequencies will be received by an observer on Earth before the lower frequencies. In these early Jodrell measurements the differences in the arrival time of the sharp leading edge of the pulses on the three frequencies could be determined with an accuracy of 5 milliseconds. If the pulse at frequency ν_1 is emitted at time T_1 and observed at time t_1 then

$$t_1 - T_1 = A + \frac{B}{\nu_1^2}$$

where A is the free space travel time and B is the dispersion constant. Fig. 24 shows the difference $(t_1 - t_2)$ plotted against $\left(\dfrac{1}{\nu_1^2} - \dfrac{1}{\nu_2^2}\right)$ for this

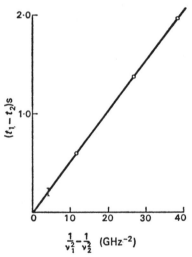

24. The dispersion in the arrival time of pulses from the pulsar PSR 1919+21 as measured at Jodrell Bank in February 1968 on frequencies of 151, 240, and 408 MHz. The point at 0·2 s delay is the earlier Cambridge observation at 80·5 and 81·5 MHz.

pulsar. The points lie on a straight line whose intercept with the vertical axis is 0 ± 5 milliseconds. Thus the differences in the emission times over the frequency band 80 to 408 MHz are either very small or they follow a similar dispersion law, which is improbable and it seemed clear from these measurements that the dispersion was occurring in the interstellar medium. The value of B derived from Fig. 24 gives a measure for the integrated electron density along the line of sight of $12 \cdot 55 \pm 0 \cdot 06$ pc cm^{-3}. The interstellar electron density is not known precisely, but the mean density was believed to be of the order of $0 \cdot 2$ electrons cm^{-3}. In this case the distance of the pulsar would be $62 \cdot 5$ parsecs.

The study of the pulse shapes in this early work indicated that on all frequencies the pulses showed a very sharp rise. For example Fig. 25 shows a sequence of 22 pulses recorded on 29 February 1968 on a frequency of 408 MHz. The mean pulse shapes averaged over 2 minutes and 12 minutes are shown in Figs. 26(a) and (b) respectively. The

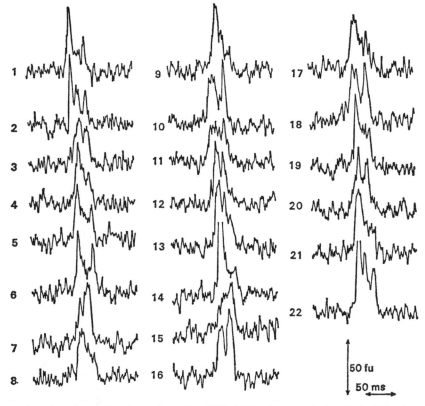

25. A series of 22 pulses from the pulsar PSR 1919+21 recorded at Jodrell Bank on 408 MHz on 29 February 1968. The receiver bandwidth was 4 MHz and the time constant 1 ms.

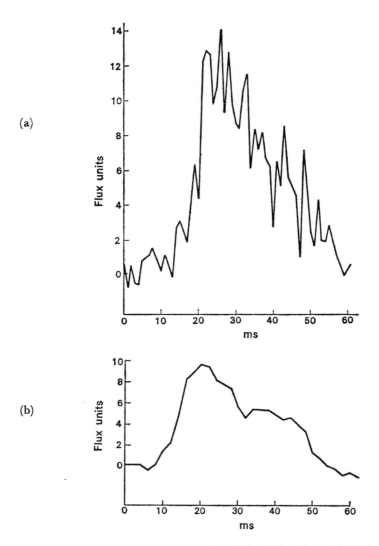

(a)

(b)

26. The mean pulse shape from the pulsar PSR 1919+21 on 408 MHz as recorded at Jodrell Bank with receiver bandwidth 1 MHz and time constant 1 ms. (a) averaged over 2 minutes at 0445 UT on 28 February 1968, (b) averaged over 12 minutes at 0650 UT on 24 February 1968.

duration of the pulse on all three frequencies was 35 milliseconds. Since the pulsar could be followed for many hours per day, it was possible to observe the behaviour of the pulse amplitude with time. On all three frequencies the amplitude was found to be variable and occasionally showed a rapid rise in amplitude by 10 times persisting for 10 minutes and then not recurring again for several hours. Fig. 27 shows the variation of pulse amplitude over a typical six-hour period. Obviously it became highly desirable to find out how the characteristics of the other three pulsating sources already discovered by the Cambridge team compared with this one. Unfortunately the Cambridge scientists had not revealed the co-ordinates of these three sources and an independent search for them would have been enormously time-consuming. The Americans too were anxious to study the phenomena and were clamouring for the

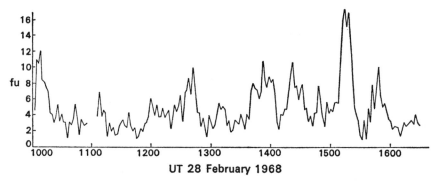

UT 28 February 1968

27. The variation of pulse amplitude from the pulsar PSR 1919+21 averaged over 2 minute intervals for an observing period of 6 hours. Recorded at Jodrell Bank on 408 MHz on 28 February 1968.

co-ordinates of the other three sources. The Cambridge people obstinately held on to the information while they carried out their own work, and the co-ordinates remained a secret. Then Hewish was persuaded to pass them on to Jodrell Bank, but only on condition that the information went no further! Fortunately the major fuss which seemed likely to develop did not materialize because the Cambridge team soon realized that it was not in their interests to hold these data and the co-ordinates were soon generally released.

For some weeks after this it seemed that all available interest and effort at Jodrell Bank was concentrated on the pulsar phenonema and there was an outpouring of results such as occurs only rarely in the life of a scientific establishment. Quickly the fire spread to many other radio observatories and there was a similar harvest of results and discoveries of new pulsars in America at Harvard, at the National Radio Observatory at Green Bank, by the Cornell Group at Arecibo, and in Australia at the Parkes Observatory of C.S.I.R.O. and at the Molonglo Radio Observatory. The theorists too got to work and within the first 6 months over 50 articles had been published in *Nature*. This may well be a record for rapidity of scientific observation and publication of the results on a single subject.

It is not possible to recount these discoveries in sequence particularly since many of the earlier results are now seen in a better perspective. It is possible only to attempt a survey of our present knowledge of the pulsars; and even this is hazardous in a topic which is still unfolding with such great speed. As far as Jodrell was concerned the excitement diminished only as the staff became exhausted. Nearly a year elapsed before pulsar observations began to take a more normal share of the observing time of the telescopes. Then one day I found both Lyne and Rickett, who had borne the brunt of the observing, working again on their theses. 'Well' I remarked to Lyne 'your thesis is going to be easy to do now, you need only include all the work you've published on pulsars during the last few months.' 'No' replied Lyne 'no pulsars, I'm concentrating entirely on the lunar occultations'!*

The discovery and nomenclature of further pulsars

The Cambridge team called their object a rapidly pulsating radio source. For some time the objects were referred to either as pulsating radio sources or pulsed radio sources. Soon, however, the word pulsar emerged from some unknown American source and this easy description soon became general usage (probably by analogy with quasar, although there is no relation between the phenomena).

When Hewish and his colleagues published the details of the three further sources which they had found by searching the original records† they suggested a nomenclature which would give a reasonably specific identification of the source in terms of its right ascension, preceded by the letters CP for Cambridge Pulsar. Thus the original source was at right ascension $19^h 19^m 38^s_1$ and wast herefore given the notation CP1919. This plan was followed when other pulsars were discovered elsewhere. Thus the first pulsar discovered by the Harvard group‡ (using the telescope at the National Radio Observatory, Green Bank, West Virginia) in June 1968 was at right ascension $15^h 06^m$ and was given the notation HP1506. Unless so many pulsars are discovered that several are found at the same right ascension, this notation which gives no indi-

---

* When Dr. Lyne read this 4 years later he commented 'To this day I do not know whether we were right in yielding to the pressures to write our theses at this time. It was because of this that we missed out in the story of the Crab.' (See Ch. 12.)

† J. D. H. Pilkington, A. Hewish, S. J. Bell, and T. W. Cole, *Nature*, **218**, 126, 1968.

‡ G. R. Huguenin, J. H. Taylor, L. E. Goad, A. Hartai, G. S. F. Orsten, and A. K. Rodman, *Nature*, **219**, 576, 1968.

cation of the declination, is satisfactory. When the Australian groups discovered pulsars in the southern hemisphere in July 1968 they did not follow this convention but proposed the contraction PSR (from Pulsar) followed by two groups of figures indicative of the right ascension *and* declination. Thus PSR 1749–28 was the pulsar they discovered at right ascension 17^h49^m and declination $-28°$. By international convention the responsibility for decisions in such matters of nomenclature lies with the International Astronomical Union and this body will no doubt comment on the situation in due course. In the meantime a tacit agreement soon developed in favour of the PSR notation.

The processes by which the Cambridge scientists discovered the original pulsars were accidental and insensitive compared with the digital techniques available once their pulsating nature was appreciated. Hence it seemed likely that large numbers of pulsars would quickly be discovered by other workers employing specifically designed search techniques. In fact, this expectation was not realized. Narrow beam search techniques which, in principle, could bring a high sensitivity to bear, had to be associated with a search in two other variables simultaneously—the pulse period and the problems associated with an unknown dispersion. The severest handicap was associated with the great variability in signal strength of the pulsars. For example, it was found that with the large steerable radio telescopes deployed in a routine scan of a selected area of the sky, the well-known and strong pulsars would frequently not be recorded when they were in the beam of the telescope. For these various reasons the total number of pulsars known to the end of 1971 was only 55. Of these 26 had been discovered in the southern hemisphere by using the large cross-type aerial at Molonglo. In the northern hemisphere at that time the discoveries had been Jodrell Bank 9, Cambridge 7, Green Bank, U.S.A. 4, Arecibo 4, Bologna 4, and U.S.S.R. 1.

Although computer techniques appear to have fundamental advantages in these searches, it is a fact that about three-quarters of the 55 pulsars mentioned above were found by examination of paper chart recordings. However, the inherent possibilities of computerized searches are clearly shown by the fact that the Mark I telescope at Jodrell Bank, although having the smallest collecting area of any of the telescopes used elsewhere in the search for pulsars, has had the greatest success rate in the northern hemisphere. Four of these Jodrell pulsars were discovered between 28 September and 12 November 1969 when engineering work was already in progress as a stage in the conversion of the Mark I to the Mark IA and a further five were discovered in May and June 1970. The presence of M. I. Large, who was primarily responsible for the pulsar discoveries using the Molonglo Cross in Australia, was a great stimulus to this work.* The technique has been described in publications by him and J. G. Davies.†

The essence of the technique was a single pulse search using an on-line computer which searched for pulses showing the frequency dispersion in arrival time which is characteristic of genuine pulsars—as distinct from random interference pulses. The two frequencies used were 406 and 410 MHz. The power in each frequency channel was digitized using voltage to frequency converters and binary counters. The Argus 400 computer was used to carry out the processing and the events were recorded on a digital graph plotter with co-ordinates of time against dispersion measure. By the time the Mk IA became available in 1972 Davies and Lyne had developed an alternative search technique so that the analysis could be carried out in real time on the Argus data processing computer. Between 26 July and 21 September 1972 they observed 3280 beam areas in a region of sky based on the galactic plane. Each beam area was observed for 11 minutes and the computer integrated and analysed the received signals for the presence of pulses in 16 000 period ranges between 0·16 to 1·45 seconds. The dispersion measures were obtained by using two bandwidths separated by 4 MHz at 408 MHz. In this search they discovered 18 new pulsars.‡

In searches of this type there are many problems militating against the rapid detection of pulsars. For example, the sensitivity is proportional to the square root of the receiver bandwidth and hence large bandwidths are needed to reveal the fainter pulsars. However, this increases the dispersion and the pulses from distant pulsars are thereby broadened and may no longer be sharp compared with the pulse repetition rate. Thus with a given bandwidth there is a lower limit to the pulse repetition rate which can be recognized. A comprehensive search for pulsars hence involves a position search in the sky and a series of searches in each sky position with varying receiver and computer recognition parameters. Of the 84 pulsars discovered in the northern and southern skies to the end of 1972,§ 27 of the 50 in the northern hemisphere had been discovered at Jodrell Bank—an illustration of the

---

* Dr. M. I. Large was one of our Lecturers (1956–63) who migrated to Australia. From September 1969 to August 1970 he returned to work with us, on study leave from the University of Sydney.

† J. G. Davies and M. I. Large, *Mon. Not. R. astr. Soc.*, **149**, 301, 1970; J. G. Davies, M. I. Large, and A. C. Pickwick, *Nature*, **227**, 1123, 1970.

‡ J. G. Davies, A. G. Lyne, and J. H. Seiradakis, *Nature*, **240**, 229, 1972.

§ The discoveries to the end of 1972 at the individual observatories total 86 as follows:

significance of steerability and advanced data handling in these difficult observations.

Cambridge, U.K.	7	Puschino, U.S.S.R.	1
Arecibo, Puerto Rico.	5	Bologna, Italy.	4
N.R.A.O., Green Bank, U.S.A.	6	Jodrell Bank, U.K.	27
Molonglo, Australia.	34	Ootacamund, India	2

Two of these pulsars were discovered independently at two observatories so that the total number of known pulsars to that date was 84.

12

The nature of the pulsars

The fluctuations in amplitude of the pulses

The original Cambridge investigations revealed that the strength of the pulses was not constant and all subsequent observations of the various pulsars showed this phenomenon. There was no obvious sign of any pattern or frequency in these fluctuations, and a number of suggestions were made about the cause of these variations. Scheuer of the Cambridge group first developed the detailed theory that they arose because of a scintillation effect caused by irregularities either in the general interstellar electron distribution or in a localized electron cloud between the Earth and the pulsar.

Although the scintillation idea seemed plausible it did not give a full explanation of the various observations over a range of frequencies. It is believed that the work of Rickett* at Jodrell Bank has now produced a fuller insight into these fluctuation phenomena. His conclusion is that there are at least two different effects which give rise to the fluctuation observed on Earth: (a) a short-term component with a timescale which is independent of the observing frequency, but which varies from pulsar to pulsar from a fraction of a second to several minutes (b) a longer-term component with a timescale which increases as the observing frequency is increased, and which varies from pulsar to pulsar (from a few minutes to several hours, on a frequency of 408 MHz).

Rickett's investigations were concerned with (b), the long-term fluctuations. The characteristics of this fading are clearly evident in Fig. 28. This is the output of the digital autocorrelation spectrometer gated to accept signals during an interval of 30 milliseconds from the pulsar PSR 0329 + 54. The observing frequency was 408 MHz and the pulse spectrum was plotted automatically at intervals of 50 seconds. In this plot a frequency width of 2·5 MHz centred on 408 MHz is displayed horizontally. The resolution on this scale is 60 kHz. Successive plots at 50 second time intervals are displayed vertically for a period of about 1 hour. This

* B. J. Rickett, *Nature*, **221**, 158, 1969; *Mon. Not. R. astr. Soc.*, **150**, 67, 1970.

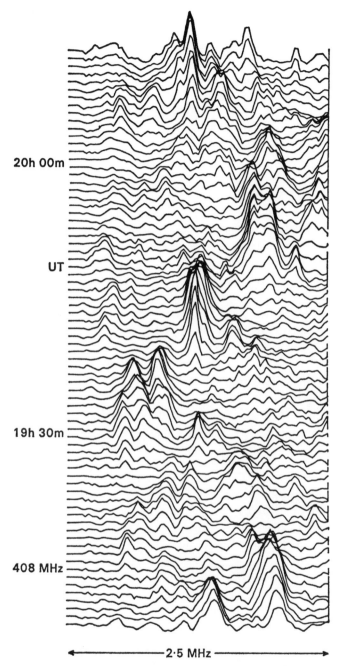

28. Successive integrations of about 70 pulses from the pulsar PSR 0329+54, plotted at 50 second intervals. The vertical scale is time and covers a period of about one hour. The horizontal scale covers 2·5 MHz in frequency centred on the observing frequency of 408 MHz.

method of display shows clearly that on this frequency the pulses from PSR 0329+54 have features with a characteristic duration of 4 minutes and a frequency width of about 150 kHz. There is no tendency for the features to drift in frequency and their variation in time and frequency seems to be random. Although the characteristics of the long-term fading are clearly displayed in this diagram, there is no indication as to whether the effects are intrinsic to the pulsar or arise as the pulses traverse the interstellar medium.

Rickett investigated this problem by studying the frequency structure of 10 pulsars on 408 MHz. He and Lyne had previously noticed* that the fading on PSR 0329+54 was much deeper if the observations were made in a receiver with a narrow bandwidth (0·35 MHz) than with a wider bandwidth (4 MHz). (The reason is, of course, evident from Fig. 28 where the characteristic feature would remain within a bandwidth of 2·5 MHz for most of the period of 1 hour there displayed.) Rickett therefore studied this effect for 11 pulsars by using a receiver with 3 different channel widths in the range 0·1 to 4 MHz, each channel being centred on the same radio frequency (408 MHz). The output from these 3 channels was fed into an on-line computer which printed out the mean pulse intensities every minute. From the characteristics exhibited in Fig. 28 it is clear that the channel with 2 MHz bandwidth will show less fluctuation than one with 0·2 MHz bandwidth. These measurements were made to determine a characteristic or half visibility bandwidth B_h for each pulsar, defined to be the bandwidth which reduces the fluctuation value by one half over that observed with an infinitely narrow bandwidth. The fluctuation value was defined in terms of the ratio of the root mean square deviation of the pulse energy to the mean pulse energy. As defined in this manner B_h is clearly several times the width of the characteristic features in Fig. 28. Rickett then plotted these values of B_h for the 11 pulsars against the integrated electron content along the line of sight to the pulsar—that is the dispersion measure $\int N.dl$. The relationship in Fig. 29 clearly shows that as the dispersion measure increases, the frequency structure becomes finer, and leads to the conclusion that the electrons between Earth and the pulsar which are responsible for the dispersion are also responsible for the frequency structure and fluctuation in intensity.

The clear inference from these measurements is that the long-term fading of the pulsar intensities is a propagation effect associated with the passage of the pulses through the interstellar medium. Theories previously developed for scintillation under these conditions predict that the points in Fig. 29 should lie on a line with a slope of −2. The agreement is

* B. J. Rickett, and A. G. Lyne, *Nature*, **219**, 1339, 1968.

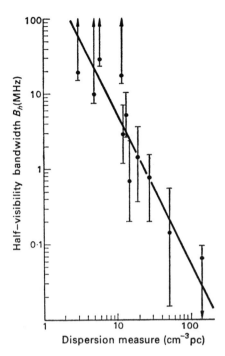

29. A logarithmic plot of the half-visibility bandwidth B_h at 408 MHz against dispersion measure for eleven pulsars. The line has a slope of -2.

close. Application of the general theory predicts that the electron irregularities in the interstellar medium have a typical size of 10^{11} to 10^{13} cm, r.m.s. deviations in electron density of 10^{-5} to 2×10^{-4} cm^{-3} and that relative transverse velocities between Earth and pulsar of 10 to 100 km s^{-1} are involved.

Although Rickett's investigations give a satisfactory explanation of the long-term fading (b) of the pulsar signals they do nothing to clarify the origin of the short-term fluctuations (a) which are independent of frequency. The latter are believed to be intrinsic to the pulsar itself, and are unlikely to be explained until a satisfactory physical description of the pulsars themselves can be given. The complexity of the phenomena observed on Earth is now seen to arise because these intrinsic short-term fluctuations are modulated by the scintillation process occurring in the transmission through the interstellar medium. The confusion was particularly severe on the first pulsar studied (PSR 1919+21) where, by chance, it happens that the two types of fading are similar in timescale and depth of modulation on a frequency of 150 MHz.

The deduction from the study of the long-term fading was that the scale size of the diffracting screen might be of the order of a million kilometres and that relative transverse velocities of the pulsar and the Earth of tens of kilometres a second were involved. This led to a joint experiment between Jodrell Bank and Penticton. The purpose of this investigation was to see if a particular fading pattern at Penticton could be identified at a later time at Jodrell Bank. The experiment was carried out by Lyne at Jodrell and Galt at Penticton on the pulsar PSR 0329+ 54. They found that a pattern at Penticton repeated at Jodrell Bank about 20 seconds later, thus indicating transverse velocities of the order of 300 to 400 km s^{-1}. It is possible that these high velocities are associated with motions in the diffracting screen, but the present tendency is to think that the pulsar itself may be moving at these high velocities. If this idea is confirmed, there are important consequences related to their possible origin in supernova explosions, and also to the length of time for which they might remain in the galaxy. For example, F. G. Smith[*] has pointed out that if a pulsar is moving 100 km s^{-1} faster than the Sun in the direction of galactic rotation, then it will escape from the galaxy in 100 million years. This is consistent with the further belief that pulsar periods lengthen with age, and that after this time of 100 million years a pulsar would be expected to have a period of more than 2 seconds. This could be related to the fact that only 3 of the 84 pulsars discovered to the end of 1972 had periods exceeding 2 seconds.

The periodicity of the pulsars

As soon as Hewish and his colleagues announced the discovery of the pulsars it was clear that the repetition rate of the pulses was constant to a remarkable degree. In their original measurements the pulses from PSR 1919+21 were displayed on the same record as the time pips from MSF Rugby. This made it possible to time the leading edge of the pulse to 0·1 s and observations over a 6 hour period gave the period between the pulses of $P_{obs} = 1·33733 \pm 0·00001$ s. These measurements were initiated when it seemed possible that the pulses might have a planetary origin and therefore the doppler effect arising because of the relative motion of the Earth and the source would be measurable as a change in this periodicity. A day-by-day measurement did, indeed, show the systematic change in period plotted in Fig. 30. A standard time was chosen (Dec. 11^{d}14^{h}01^{m}00^s UT)—this corresponded to the centre of the reception pattern of the aerial—and the time interval T was obtained by daily measurements of the interval between this standard time and the pulse immediately following. After 11 December the subsequent standard times were at the sidereal day intervals (23^{h}56^{m}04^s). If the recurrence period of the pulses was constant, then Fig. 30 should show a

[*] F. G. Smith, *Rep. Progr. Phys.*, **35**, 399, 1972.

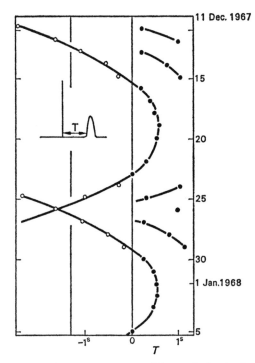

30. The original Cambridge measurements of the variation in the day-to-day pulse arrival time from the pulsar PSR 1919+21.

linear increase or decrease in T. It fact there was a marked curvature in the sense of a steadily increasing frequency.

This curvature was shown to arise because of the doppler effect occasioned by the Earth's motion alone, by means of the following argument. If N_0 is the number of pulses emitted per day by the source, V the orbital velocity of the Earth, φ the ecliptic latitude of the source and n an arbitrary day number ($n = 0$ on 1968 Jan. 17 when the Earth has zero velocity along the line of sight to the source) then the number of pulses received per day on Earth is given by,

$$N = N_0\left(1 + \frac{V}{c}\cos\varphi . \sin\frac{2\pi n}{366 \cdot 25}\right)$$

The values of n for which $\dfrac{\delta T}{\delta n}$ is zero, corresponding to an exactly integral value of N occur at $n_1 = 15 \cdot 8 \pm 0 \cdot 1$ and $n_2 = 28 \cdot 7 \pm 0 \cdot 1$ N increases by 1 pulse between n_1 and n_2. Hence the above equation becomes

$$1 = \frac{N_0 V}{c}\cos\varphi\left[\sin\frac{2\pi n_2}{366 \cdot 25} - \sin\frac{2\pi n_1}{366 \cdot 25}\right]$$

giving $\varphi = 43° 36' \pm 30'$ equivalent to a source declination of $21° 28' \pm 30'$ which, within the accuracies of the measurement and the formulae, agreed with the measured declination of the pulsar.

The conclusion that the variation in periodicity was a doppler effect, arising solely because of the motion of the Earth, was critical in removing the possibility that the source was in the planetary system. Furthermore, by making allowance for the doppler shift the true (heliocentric) period of the pulsar was derived as

$$P_0 = 1{\cdot}3372795 \pm 0{\cdot}0000020 \text{ s}$$

In fact, it quickly transpired that this period was incorrect. In a communication received by *Nature* on 16 April 1968* observers using the Parkes telescope in Sydney published a value for the periodicity of

$$P_0 = 1{\cdot}33730109 \pm 7 \times 10^{-8} \text{ s}$$

in disagreement with the Cambridge value by an amount well in excess of the combined errors. Another communication to *Nature* received on 6 May 1968† from Arecibo confirmed this Parkes value for P_0. The explanation was that in deriving their value for P_0 the Cambridge team counted one pulse too many per sidereal day.

The periodicities of the first 4 pulsars discovered in the Cambridge investigations were in the range 0·253 to 1·337 seconds. With the discovery of additional pulsars this periodicity range was extended but the periodicities of the 84 pulsars at present known are all within the range from 33 ms to 3·7 s. The constancy in the period of the individual pulsars was remarkable. The high stability of the period is naturally of great interest in connection with the physical nature of the sources and other important applications were soon pointed out—for example, their use as astronomical clocks for testing the theory of general relativity which predicts that terrestrial clocks should go more slowly at perihelion than at aphelion (because of the change in the solar gravitational potential). Since the heliocentric periods of the pulsars were determined almost immediately to 1 pt in 10^7 it seemed possible to improve this to 1 pt in 10^9 by observations over a year, and these results would approach the accuracy needed for the relativity test. It was, therefore, hardly surprising that those observers equipped with the necessary apparatus gave close attention to the systematic timing of the pulsar emissions.

Precise timings had been made at Jodrell Bank since 25 February on PSR 1919+21 and since 9 March on PSR 0834+06, 0950+08, and 1133+16. During the autumn of 1968 it became clear that all of these

---

* V. Radhakrishnan, M. M. Komesarov, and D. J. Cooke, *Nature*, **218**, 229, 1968.
† G. Zeissig and D. W. Richards, *Nature*, **218**, 1037, 1968.

pulsars were slowing down. As soon as the evidence became incontro-vertible the results were prepared for publication in *Nature*. When the paper was in draft, Graham Smith extracted the news from Cambridge that they had also found the same effect and were proposing to publish. In the event neither Jodrell Bank nor Cambridge were first with the news, since before these communications were despatched an I.A.U. Circular* arrived with the information that the Cornell group had found that the pulsar PSR 0531+21 near the Crab Nebula was slowing by one part in 2000 per annum. Stimulated by this, Graham Smith announced the Cambridge and Jodrell results at the meeting of the Royal Astro-nomical Society on 13 December 1968. The scientific papers from Jodrell and Cambridge were finally published in *Nature* on 4 January 1969.†

In order to establish the existence of this slowing down of the pulsars with certainty, precision measurements over many months are neces-sary to exclude the possibility that an error in the assumed position of the pulsar is giving rise to a spurious effect. At Jodrell Bank the pulse arrival times were measured to an accuracy of better than 2 milliseconds. The arrival time has to be corrected to the centre of the Earth and then to the barycentre of the solar system in order to remove the doppler effect arising from the Earth's motion. To do this the position of the pulsar must be known, and if this correction is in error, there will be a sinusoidal timing error with a period of 1 year. The difference Δt between the actual arrival times and the predicted arrival times for the 4 pulsars measured are shown as a function of date in Fig. 31. If the change of Δt with n was linear, this would merely indicate that the assumed period was incorrect. In fact, three of the curves in Fig. 31 are well fitted by a parabola and the interpretation is that the period is changing linearly with time—since then, $\Delta t \propto n^2$. In the case of PSR 1919+21 the observed points show an apparent sinusoidal varia-tion on the best fitting parabola and this indicates that the assumed position of that pulsar was in error by 15 seconds of arc. The rates of slowing down for the 4 pulsars obtained from Fig. 31 were

PSR 0834+06	124	ns per year‡
PSR 0950+08	38·5	,,
PSR 1133+16	110	,,
PSR 1919+21	25·0	,,

* D. Richards, I.A.U. Circular No. 2114 (1968).

† J. G. Davies, G. C. Hunt, and F. G. Smith, *Nature*, **221**, 27, 1969; T. W. Cole, *Nature*, **221**, 29, 1969.

‡ ns = nano second = 10^{-9} second.

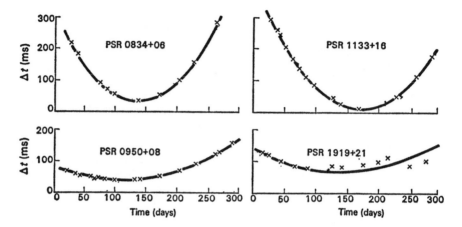

31. The difference Δt between the measured and predicted arrival times for 4 pulsars. The initial date is Julian Day 2439900·5 (14 February 1968).

The Cambridge results were similar for PSR 0834+06, 1133+16, and 1919+21. They did not measure PSR 0950+08 but studied PSR 0809+74 for which they found no change greater than 10 ns per year.

It is generally agreed that these changes are the result of physical alterations in the pulsar and if the assumption is made that the change has been linear since the object became a pulsar, then the lifetime of the Crab Nebula source, PSR 0531+21, would be 2400 years, but the lifetime of the others would be much greater—of the order of 1 to 100 million years.

Just as astronomers were becoming accustomed to this concept of the gradual slowing down of the repetition rate of the pulsars, another great surprise occurred in the spring of 1969. One of the pulsars discovered by the Australian scientists at the Molonglo Radio Observatory lay close to the supernova remnant Vela X and an association of this pulsar, PSR 0833—45, was tentatively made with the remnant.* When the pulsar similarly associated with the Crab supernova was found to be slowing down, the observers with the 210 ft Parkes radio telescope turned their attention to PSR 0833—45 and by observations over a 12 day period in December 1968 they established conclusively that this was also slowing down.† Their results (shown in Fig. 32) indicated that

* M. I. Large, A. E. Vaughan, and B. Y. Mills, *Nature*, **220**, 340, 1968.
† V. Radhakrishnan, D. J. Cooke, M. M. Komesaroff, and D. Morris, *Nature*, **221**, 443, 1969.

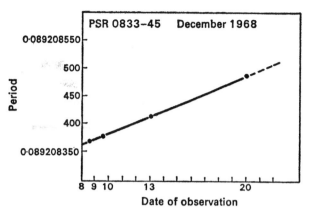

32. The variation of the period of pulsar PSR 0833–45 over a 12 day interval in December 1968. The slope corresponds to an increase in the period of 10 ns or 1 part in 10^7 per day.

the period of the pulsar is increasing at the rate of 10 ns per day or 1 part in 24 000 per year.*

These timing observations were continued at Parkes on 19 and 20 February 1969 and over this longer interval the result was refined to an increase in period of $10 \cdot 69 \pm 0 \cdot 2$ ns per day between 8 December 1968 and 20 February 1969. At the next timing run in mid-March the results were found to be entirely different—apparently the pulse rate had speeded up some time between 20 February and March 13 by 196 ns and had then continued to decrease at the normal rate. This remarkable result is shown in Fig. 33.

Fortunately, scientists at the Goldstone Tracking Station of the Jet Propulsion Laboratory, using the radio telescope normally employed for tracking space probes, were observing this pulsar, on a weekly basis and were able to tie down this sudden change more precisely.† On 24 February their results indicated that the pulsar was behaving normally but at the next observation on 3 March the discontinuity was evident. These American results are shown in Fig. 34. At some time between 24 February and 3 March this dramatic change occurred, but since no one has data between these periods the detailed history of the change remains obscure. The Australians concluded that the rate of change in

* That is on the assumption that the change has been linear, then the lifetime for the Vela X pulsar would be about 24 000 years. The pulsar PSR 0611+22 discovered in the search at Jodrell Bank referred to on p. 138 lies near the supernova remnant IC 443 and the lifetime of the pulsar deduced from the change of period is 124 000 years. These results are of the same order as the ages suggested for the supernova remnants from optical studies.

† P. E. Reichley and G. S. Downs, *Nature*, **222**, 229, 1969.

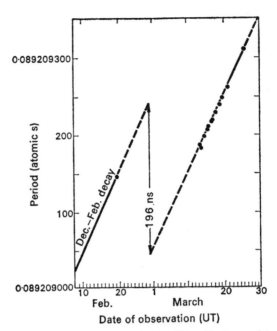

33. The heliocentric period of the pulsar PSR 0833—45 observed in February and March 1969. At some time between 19 February and 13 March the period decreased by 196 ns (see Fig. 34).

period after the sudden discontinuity was the same as before. However, the American timings were more precise, and it appears that before 24 February the rate of change of period per period ($\Delta P/P$) was

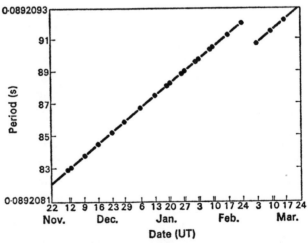

34. The barycentric period of PSR 0833—45 observed from 22 November 1968 to 24 March 1969 showing the 134 ns decrease between 24 February and 3 March.

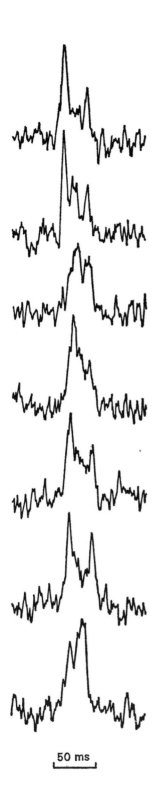

(a)

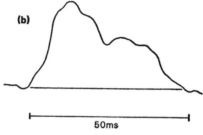

(b)

50ms

35(a) 7 consecutive pulses recorded on 408 MHz from the pulsar PSR 1919+21 (bandwidth 4 MHz, time constant 1 ms) and
(b) the mean pulse profile from PSR 1919+21 averaged over 8 minutes (bandwidth 1 MHz, time constant 2 ms).

50 ms

1.242619×10^{-13} but after 3 March it was 1.25264×10^{-13}. The discontinuous speeding up between 24 February and 3 March was 208 ns* and if no further sudden discontinuities occurred this effect would be wiped out in 6·6 years; by which time the pulsar would have the same period as it would have attained at the old decay rate, if the discontinuity had not occurred. Thirty-one months later another similar discontinuity was observed in the period of this pulsar. The observation o these extraordinary events injected still further problems into the concept of pulsars as spinning neutron stars.

The pulse shapes and polarization

In the original Cambridge paper reporting the discovery of the pulsars, the signals (from PSR 1919+21) were described as consisting of a series of pulses each lasting about 0·3 s. Immediately after the discovery was announced more sophisticated high-speed recording techniques were used, and the duration of the pulses was found to be in the region of 10 to 50 ms for the first group of pulsars. However, on 27 April 1968, Lyne and Rickett communicated to *Nature*† the results of an investigation into the pulse shapes using fast galvanometer recordings. They found that the individual pulses were variable from pulse to pulse and frequently contained features which were unresolved by a time constant of 1 ms. The mean pulse integrated over several minutes is a composite of these individual features, which may occur at differing phases within the mean envelope. This conclusion which has subsequently been found to apply to all the pulsars studied, is well illustrated in Fig. 35 which shows (a) 7 consecutive individual pulses from PSR 1919+21 and (b) the mean pulse profile from PSR 1919+21 integrated over 8 minutes. The mean pulse shape appears to be characteristic of the pulsar. The envelopes of the mean pulse for a number of pulsars is shown in Fig. 36. If the averaging is carried out for a sufficient time to smooth out the rapidly changing fine structure, these characteristic envelopes are stable and repeatable. In the case of PSR 0950+08 Lyne and Rickett‡ discovered an 'interpulse' about half way between the main pulses. A similar interpulse was found subsequently in the pulsar PSR 0531+21.

An important question concerns the polarization of these pulses. The search for linear polarization by Lyne and Smith§ was successful. They

* That is the difference between the rate up to 24 Feb. extrapolated to March 3 compared with the period measured on 3 March. This is, of course, greater than the 134 ns difference in Fig. 34 observed between 24 Feb. and 3 March.

† A. G. Lyne and B. J. Rickett, *Nature*, **218**, 326, 1968.

‡ A. G. Lyne and B. J. Rickett, *Nature*, **218**, 934, 1968.

§ A. G. Lyne and F. G. Smith, *Nature*, **218**, 124, 1968.

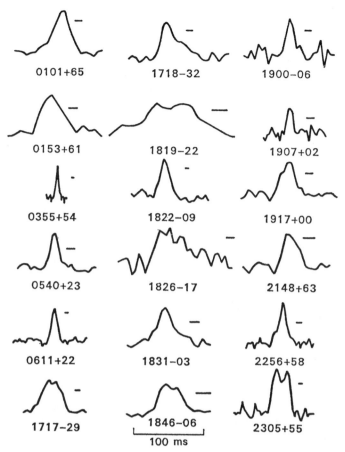

36. The mean pulse shapes of the 18 new pulsars discovered with the Mk IA telescope in 1972, obtained by integrating the total intensity of the individual pulses for periods of 10 minutes to 1 hour. The thick horizontal bars represent the instrumental resolution appropriate to each source. The timescale, which is the same in all cases is given by the 100 ms bar at the bottom of the figure.

used orthogonal dipoles on the Mk I radio telescope at Jodrell Bank on a frequency of 151 MHz with fast galvanometer recordings to study the individual pulses. A high degree of linear polarization was found for the 4 pulsars studied, with a rotation of the angle of polarization consistent with Faraday rotation in the ionosphere. However, the phenomenon of polarization was soon shown to be extraordinarily complex, and the first definitive measurements of the polarization phenomena were made by Clark and Smith* at Jodrell Bank by using a technique which displayed the polarization parameters of the individual pulses. In these observations continuous recordings were made on film with a

* R. R. Clark and F. G. Smith, *Nature*, **221**, 724, 1969.

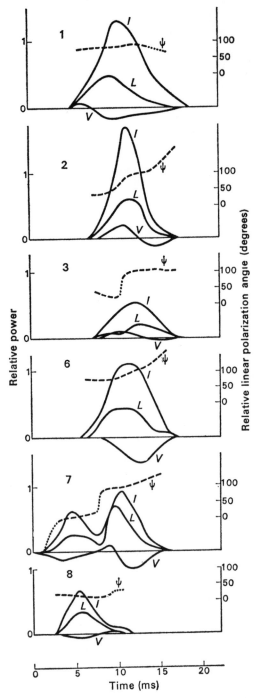

37. Analysis of six out of eight successive pulses from the pulsar PSR 0329+54 showing the complex polarization phenomena. Q and U are linear components and $L = (Q^2+U^2)^{\frac{1}{2}}$. V is the circularly polarized component (right minus left-hand polarization), ψ is the position angle $\tan^{-1}(Q/U)$. I is the total intensity. In this sequence pulses 4 and 5 were missing on the film record.

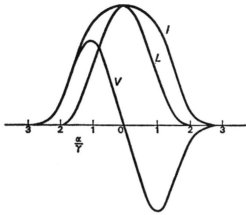

38. The theoretical prediction of the polarization parameters equivalent to Fig. 37 for synchrotron radiation from a collimated beam of electrons, as a function of the angle a between the line of sight and the orbital plane of the electrons. The angle a is expressed in terms of γ^{-1} where the electron energy is $\gamma m_0 c^2$.

camera mounted on a four-channel oscilloscope whose time base was triggered at a fixed phase of the pulsar signal. The 4 traces of the film were related to the polarization parameters. Fig. 37 shows the analysis of 6 out of 8 successive pulses from PSR 0329+54 and reveals the complexity of the phenomena. An outstanding characteristic is that there is a circularly polarized component V which changes hands during a single pulse. The position angle ψ of the linearly polarized component L, is constant or varies slowly across the pulse. The authors pointed out the striking similarity between these records and the theoretical polarization parameters expected for the synchrotron radiation from a collimated beam of electrons, shown in Fig. 38—an important result to be explained in any theoretical model of the pulsars.

The distance of the pulsars

The original Cambridge measurements showed that the pulsar PSR 1919+21 had no measurable parallax; or at least no parallax greater than 2 minutes of arc, which implied that its distance was at least 1000 astronomical units. This placed it far beyond the solar system. The next step was to establish an upper limit to the distance. As soon as it was established that the frequency sweep was a dispersive effect then, assuming a value for the interstellar electron density N, an estimate of the distance L could be obtained from the rate of frequency drift dv/dt by using the formula on p. 132. Taking a mean value for N of 0·2 electrons cm^{-3} and from the measured rate of drift dv/dt of $-4\cdot9$ MHz s^{-1}, the distance of PSR 1919+21 was estimated to be 65 parsecs. Even allowing for uncertainties in the value to be taken for N, it was immediately evident that this pulsar, and the others discovered shortly afterwards, were galactic objects. Subsequently a value for N of 0.1 electrons cm^{-3} has been assumed and the various distances calculated on this assumption from the measured rate of frequency drift are in the range of 100 to 1500 parsecs.

Although these measurements established without much doubt that the pulsars were objects within the local galaxy, the uncertainty in the value assigned for the interstellar electron density N, still left open the question of whether they were in the solar neighbourhood or in the distant regions of the galaxy. The development of an entirely different method of measuring the distance of pulsars was therefore a matter of great interest, not only because of the distance measurement itself but also because the formulae on p. 132 for the dispersion could then be used to obtain a more realistic value for N. The technique which was developed at Jodrell Bank in the autumn of 1968* depended on a measurement of the absorption of the pulsar radiation in the intervening clouds of neutral hydrogen. In a stationary system this absorption would occur at the precise hydrogen line wavelength of 21 cm. However, the rotation of the galaxy imposes a doppler shift. The extensive measurements of the observed frequency of the neutral hydrogen emission from various directions, since the discovery of the H line emission in 1951, has enabled astronomers to establish reliable models of the position and movement of the large clouds of neutral hydrogen in the galaxy. Thus the actual frequency at which the pulsar radiation is absorbed establishes at once, on the basis of these models, the neutral hydrogen cloud system through which the radiation has passed.

This measurement poses an exceedingly difficult technical problem, firstly, because of the intermittent nature of the emission from the pulsar, and secondly, because the pulsar emission is itself rather weak at these high H line frequencies. The autocorrelation digital spectrometer was used with a resolution of 14 kHz over a band of 560 kHz. This compared the spectrum of the radiation in the telescope beam during the pulse with that in the intervals between the pulses. An indication of the problem is given by the estimate that the mean power from the pulses was only 10^{-4} of the total power in the receiver passband.

The pulsar PSR 0329+54 was chosen for the experiment since this had measurable emission in the 1420 MHz region of the spectrum, and also lay close to the galactic equator in the direction of the Perseus arm, so that the emission from it might be anticipated to pass through hydrogen clouds of sufficient depth to exhibit the absorption. The final

* G. de Jager, A. G. Lyne, L. Pointon, and J. E. B. Ponsonby, *Nature*, **220**, 128, 1968.

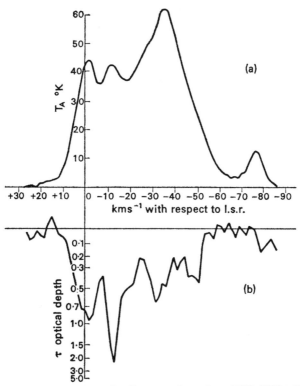

39. The first attempt to measure the distance of a pulsar (PSR 0329+54) at Jodrell Bank in 1968 by studying the neutral hydrogen absorption. (a) shows the normal neutral hydrogen line emission in the direction of this pulsar and (b) shows the neutral hydrogen absorption in the spectrum of the pulsar.

results of this analysis are shown in Fig. 39: (a) shows the normal neutral hydrogen emission spectrum in the direction of the pulsar; the pulsar absorption spectrum is aligned with this in (b). The correspondence between the main features of these two spectra indicates that the pulsar must lie beyond hydrogen clouds having velocities of -58 km s^{-1} with respect to the local standard of rest. According to the generally accepted galactic rotation model these clouds must lie at 4·2 kpc from the Sun on the far side of the Perseus arm. The figures also show a possible corresponding feature at -80 km s^{-1} which, if established, would place the pulsar beyond hydrogen clouds in the outer galactic arm at a distance of more than 6 kpc from the Sun.

This was a surprising result. The dispersion measurements gave a total electron content along the line of sight of 26·75 pc cm^{-3}. Thus even at the lower limit of the distances of 4·2 kpc, the electron density N is calculated to be only 0·006 cm^{-3}, or 17 times less than the value of N used in estimating previous pulsar distances. Although this density was so low compared with the values for N previously assumed there was other evidence published at the same time* based on measurements of the low energy cosmic ray flux in the solar neighbourhood which indicated a similar value for N.

It soon transpired that the distance of PSR 0329+54 derived in the Jodrell Bank measurements was too large. The experiment was repeated with a more extensive frequency coverage in order to assess the validity of the feature at -80 km s^{-1}. It was then found that the baseline had been incorrectly established in the original work and that Fig. 39 did, in fact, only show the first absorption feature. This placed PSR 0329+54 in or near the first spiral arm at a distance of the order of only 1000–2000 pc. At the same time observers at Nançay in France and at Green Bank in America repeated the Jodrell Bank technique and published this lower value for the distance. In the extreme heat of international competition which developed in the pulsar research at that stage, the desire and ability to publish results instantly had the drawback that a number of errors were made. The Jodrell estimate of the distance for this pulsar was one such error. Nevertheless, a new technique had been established, and had at least produced independent evidence for the order of magnitude of the distance of the pulsars. Subsequently the Nançay team used the technique to measure the distance of two other pulsars PSR 2015+28, which they found to be at a distance greater than 1000 pc, and PSR 1933+16 at 6000 pc.

Optical identification of the pulsars

Immediately the discovery of the pulsars was announced strenuous efforts were made to identify them with photographic objects. Naturally the Cambridge team, with prior knowledge of the discovery, were in a strong position and Ryle and Bailey[†] soon published a suggested identification of the first pulsar PSR 1919+21 with a blue star of 18th magnitude on the Palomar Sky Survey. Neither this identification, nor those suggested for other pulsars, survived the impact of more accurate information. Neither did any of the attempts made to detect optical flashes from the regions of the radio pulsars meet with any significant success, although there were false alarms.

In the autumn of 1968 the Australian scientists[‡] using the Molonglo radio telescope discovered a pulsar with the shortest period then known (0·089 s). The position of this pulsar (PSR 0833−45) is close to that of the Vela X supernova remnant. It was suggested that this association, and its extremely short period, supported the idea that the pulsar was a

* H. J. Habing and S. R. Pottasch, *Nature*, **219**, 1137, 1968.
† M. Ryle and J. A. Bailey, *Nature*, **217**, 907, 1968.
‡ M. I. Large, A. E. Vaughan, and B. Y. Mills, *Nature*, **220**, 340, 1968.

rotating neutron star and was the remnant of the supernova explosion. Furthermore it was suggested that over long periods of time the transfer of angular momentum from the star to the interstellar medium would cause slowing down of the rotation and a lengthening of the observed period, which was indeed later found to be happening. This pulsar PSR 0833−45 again attracted attention when the abrupt decrease in its period was observed in February 1969 (see page 148).

The evidence for a possible association of pulsars with supernova remnants soon received further strong support when the Americans, using the 300 ft transit radio telescope at Green Bank discovered two pulsars near the Crab Nebula in October 1968*. These were originally designated NP 0527 and NP 0532 (later PSR 0525+21 and PSR 0531+21) and it was suggested that these pulsars were associated with the Crab Nebula. It was the latter of these two pulsars which soon attracted close attention when observations at Arecibo† showed that its position was indeed within 10 minutes of arc of the centre of the Crab nebula and also that it had the shortest period of any known pulsar: 33 ms. This was even shorter than that of the Vela X pulsar of 89 ms. The main pulse was followed by another 14 ms later, with about one quarter of the intensity of the main pulse. With regard to the first of these pulsars, PSR 0525+21 later measurements giving a more accurate position showed that it was 1·5 degrees from the centre of the nebula and also had the longest period of any known pulsar (3·7 s). Thus it appears exceedingly unlikely that PSR 0525+21 is related either to the Crab nebula or to PSR 0531+21.

In the case of PSR 0531+21 the evidence for its association with the Crab nebula quickly and dramatically become conclusive. On Saturday morning, 18 January 1969, the secretary on duty at Jodrell Bank handed me the following message which had just appeared on the telex machine:

NP 0532 OBSERVED OPTICALLY JANUARY 15-16 FIVE SECONDS NORTH FOUR SECONDS EAST OF SOUTH PRECEDING STAR OF CRAB CENTRAL DOUBLE ESTIMATED ERROR FIVE SECONDS GEOCENTRIC PERIOD 33095 MICROSECONDS WIDTH FOUR MILLISECONDS OCCASIONAL SECONDARY PULSES APPROXIMATELY MIDWAY BETWEEN PRIMARIES INTEGRATED VISUAL MAGNITUDE EIGHTEEN PEAK MAGNITUDE FIFTEEN COCKE DISNEY TAYLOR STEWARD OBSERVATORY. MARSDEN.

* D. H. Staelin and E. C. Reifenstein, I.A.U. Circular No. 2110, 1968, and *Science*, **162**, 1481, 1968.

† R. B. E. Lovelace, J. M. Sutton, and H. D. Craft, I.A.U. Circular No. 2113, 1968.

A few days later the I.A.U. circular arrived* with the additional information that the observations had been made on the 91 cm reflector, and almost immediately another circular† came confirming that the light flashes from PSR 0531+21 had been found also at the McDonald Observatory and at Kitt Peak on 20 January. The coincidence of the periods and positions of the optical and radio pulses left no doubt that at last a definite identification of a pulsar had been achieved. Fig. 40 shows the appearance on the recorder of the first optical pulse observed by the Arizona astronomers on 16 January. Fig. 41 shows the effect of summing 300 pulses. In addition there was a smaller interpulse as found in the radio spectrum.

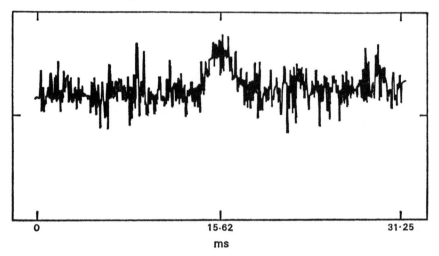

40. The first optical pulse observed on the recording screen of the 36 inch telescope at the Steward Observatory (Arizona) 16 January 1969, 03ʰ30ᵐ UT from the Crab nebula pulsar. The amplitude scale is arbitrary, and represents the superposition of about 5000 optical pulses.

Not the least remarkable feature of this observation was that the radio pulses and the light flashes coincided with the star in the Crab nebulosity which twenty-seven years previously Baade‡ and Minkowski,§ in their classical study of the Crab, had suspected was the remnant of the supernova explosion producing the nebula. These initial results were

   * I.A.U. Circular No. 2128 and subsequently W. J. Cocke, M. J. Disney, D. J. Taylor, *Nature*, **221**, 525, 1969.
   † I.A.U. Circular No. 2129 and subsequently R. E. Nather, B. Warner, M. MacFarlane, *Nature*, 221, 527, 1969. Also R. Lynds, S. P. Maran, D. E. Trumbo, *Astrophys. J.*, **155**, L121, 1969.
   ‡ W. Baade, *Astrophys. J.*, **96**, 188, 1942.
   § R. Minkowski, *Astrophys.J.*, **96**, 199, 1942.

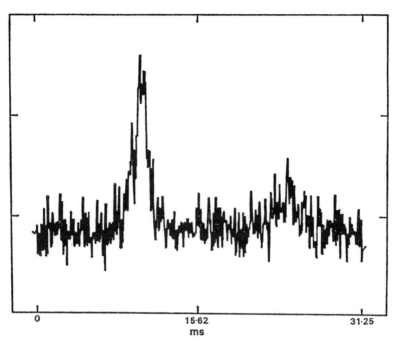

| 0 | 15·62 | 31·25 |

ms

41. The effect of summing 300 optical pulses from the Crab nebula through a smaller diaphragm. (5 arc s, instead of the 22 arc s diaphragm used to obtain Fig. 40.) 20 January 1969, 06ʰ00ᵐ UT.

obtained by using a conventional photoelectric photometer, with the output fed into a multiscaler analyser operating near the pulse periods. Even more dramatic results were achieved on 3 February at the Lick Observatory, when a television camera with an image intensifier was used on the 120 inch telescope.* The light beam from the telescope was chopped by a rotating disc so that the 'open' periods were separated by a time close to that of the period of the optical flashes. The picture frames were recorded on video tape and the analysis of these is shown in Plate 3. Apparently the major part of the light received from the star is actually the integrated effect of the light flashes.

A sad footnote for British astronomers is that after the announcement of this discovery, astronomers at the Cambridge Observatories analysed records taken on 24 November 1968 in which they were also searching for light flashes from the Crab coincident with the radio pulses from PSR 0531+21. They found that they had, in fact, observed the light flashes in November but had not analysed the data previously 'because no flashes had been detected from the many pulsars studied at different

* J. S. Miller and E. J. Wampler, *Nature*, **221**, 1037, 1969.

observatories with sensitive equipment and because the radio position available on 24 November was uncertain'.*

The exciting saga about PSR 0531+21 did not end with the discovery of the light flashes. On 13 March H. Friedman† of the U.S. Naval Research Laboratory, launched a rocket containing X-ray detection apparatus for pulsed observations, and discovered that 8 per cent of the total X-ray flux from the Crab appeared in the form of pulses with double structure as in the radio and optical case. Eight days later scientists of Columbia University announced that they had, in fact, observed the X-ray pulses on 7 March in an Aerobee rocket launched from White Sands which pointed towards the Crab nebula for 240 seconds.‡ No other pulsar has yet been associated with similar X-ray pulsation. During the last few years X-ray observations from Earth satellites have revealed a number of pulsating X-ray sources (for example in Cygnus and Scorpius), but the current tendency is to associate these with binary systems and not with pulsars since no radio pulses have been detected.

In 1969 and 1970 there was a further remarkable discovery concerning the Crab nebula pulsar. γ-ray pulses in phase with the optical and X-ray pulses (including the interpulse) were discovered. On 25 October 1969 balloon-borne equipment for the detection of γ-rays in the energy range 10–150 MeV was launched from Texas by scientists of the Naval Research Laboratory, Washington D.C.§ The γ-rays were detected through their interaction with a nuclear emulsion stack. The electrons produced by this interaction traversed a multi-wire proportional counter and plastic scintillation counters. When a γ-ray was detected, a spark chamber of two 4 inch gaps was actuated and recorded stereoscopically on a 35 mm film. Subsequently, after recovery of the equipment, the electron tracks in the nuclear emulsion were located from measurements on the sparks. The tracks could then be followed back to the γ-ray interaction and by this procedure the direction of the incident γ-rays could be determined to within about 2 degrees and their energies to about 25 per cent.

A comparison of the distribution of events within 10 degrees of the Crab nebula with those more than 10 degrees away, showed a high probability that the nebula was emitting pulsed γ-radiation of energy greater than 10 MeV in phase with the optical and radio pulses. Furthermore, the measurements indicated that this pulsed radiation dominated

* R. V. Willstrop, *Nature*, **221**, 1023, 1969.
  † I.A.U. Circular No. 2141, 22 April 1969.
  ‡ J. R. P. Angel, R. Novick, P. Vanden Bout, M. Weisskopf, R. Wolff, I.A.U. Circular No. 2142, 30 April 1969.
  § R. L. Kinzer, R. C. Noggle, N. Seeman, and G. H. Share, *Nature*, **229**, 187, 1971.

the γ-ray emission (in the X-ray energy range only about one-fifth to one-tenth of the X-rays are in the form of the pulses). Although these measurements were made on 25 October 1969, the results were not published until 15 January 1971. In the meantime a group of British scientists from the University of Bristol had published their results in December 1970 of γ-ray measurements from a balloon using a different technique and had confirmed the existence of these pulses.* The balloon was launched from Bedford on 2 August 1970, and they established the existence of the pulsed γ-radiation in phase with the optical and radio pulses. Their results also indicated that the pulsed γ-radiation dominated at photon energies greater than a few MeV.

The pulsed emission from the Crab nebula pulsar has thus been detected from the radio to the γ-ray region of the spectrum, spanning a frequency range from about 3×10^7 Hz at the lowest radio frequency to 3×10^{20} Hz at the highest γ-ray energy—a wavelength range of 10^{13} to 1. Amongst the pulsars discovered so far this Crab nebula pulsar is certainly unique. It is the only object from which in-phase optical, radio, X-ray, and γ-ray pulses have been detected. Indeed, it is the only pulsar which has been identified by other than radio pulses. Although the association of the pulsar PSR 0833 — 45 with the Vela nebula in the southern hemisphere seems highly probable, no optical pulses have been detected from this object. The unique properties of the Crab nebula stimulated the organization of an international symposium at Jodrell Bank in August 1970 to discuss this single object.†

What are the pulsars?

As soon as the short duration of the pulses was established one incontrovertible fact about the pulsars became obvious: they must be exceedingly small bodies by astronomical standards. This is because a radiating body cannot emit a pulse of duration shorter than the travel-time of radiation across it. This means, for example, that a pulse of 10 milliseconds duration cannot arise from a radiating body whose radius is greater than 3000 km. From the measured pulse durations it is clear that the known pulsars must have sizes of this order, that is, less than the size of the Earth and planets.

Two other critical features were also immediately obvious: first, that although the pulsars were of planetary dimensions, they were at stellar distances; second, that the pulses were emitted with extreme regularity. The first point indicated that pulsars were probably extremely small and

---

* R. R. Hillier, W. R. Jackson, A. Murray, R. M. Redfern, and R. G. Sale, *Astrophys. J.*, **162**, L177, 1970.

† *The Crab Nebula* (D. Reidel Publishing Co., The Netherlands, 1971).

dense stars, and the second that the emission was associated with some gravitational control such as rotation or vibration, rather than with events in the atmosphere of the object. In such cases there is a limit to the shortness of the period determined by the density of the body ρ and the constant of gravitation G. The order of magnitude of the shortest possible period is inversely proportional to $\sqrt{G\rho}$. Thus, for example, with a typical pulsar period of 0·5 s a density of 10^8 g cm^{-3} is involved. This is the order of density which would apply if the mass of a typical star were contained within a volume of planetary dimensions, hence there is an overall consistency if one postulates that the pulsars are exceedingly dense stellar objects in which the star's mass is contained within a few thousands of kilometres (instead of within a million km as in the case of the Sun) and that the pulses are associated in some manner with the rotation, oscillation, vibration, or other mechanical feature of the star.

Stars of this type are known—the white dwarfs. They are believed to have reached the final stages of their evolutionary sequence in which the nuclear fuel has been expended, and there is insufficient thermal pressure to prevent the gravitational collapse of the star. Thus it was natural to think in terms of the white dwarfs as sources of the radio pulses. However, none of the pulsars coincided with any of the known white dwarf stars, and furthermore the density in the white dwarfs, although exceedingly high, was still one hundred times too small to explain the observed periods according to the general arguments given above.

Astrophysicists had for a long time speculated that the white dwarfs might not be the end product of stellar evolution, but that under the unstable conditions which exist in a white dwarf, processes might occur which would cause the protons and electrons to combine to form neutrons. In such cases a further collapse would occur as the electron pressure vanished and eventually the entire stellar material would consist of neutrons and a stable condition would not be reached until the kinetic pressure of these neutrons balanced the gravitational forces. Calculations had indicated that such 'neutron stars' would have densities of the order of 10^{14} g cm^{-3} and that their dimensions would be of the order of only 100 km or less. Furthermore there was already an extensive theoretical literature dealing with the pulsation of neutron stars as end products of supernova explosions. These general considerations led Hewish and his colleagues to suggest, in their first publication, that the pulsars were neutron stars.

There were however difficulties, which for sometime seemed insurmountable. Whereas white dwarfs were not dense enough to give the observed pulsar periods, the neutron stars were far too dense. If the pulsations were arising from objects with the enormous density of neutron stars, the periods should be only a millisecond—and the observed periods were of the order of a second. It is hardly surprising that the literature soon became full of a variety of theories to explain the pulsars. Then, with the discovery that the periods of some of the pulsars were much shorter, and specifically that the pulsar in the Crab nebulosity was not only of short period but also slowing down, it became reasonable to suggest that millisecond periods might well apply in the early stages of pulsars. If this phase was associated with the epoch of the supernova event, and the formation of the neutron star as a remnant, then the chance of finding pulsars with periods of only a few milliseconds would be extremely small and—as indeed the case has proved—the observed periods would be much longer.

T. Gold* of Cornell was the first to suggest that the rotation of a neutron star would provide the type of astronomical clockwork event which seemed to be required. He predicted the change in period with time, since the rotation would gradually slow down as energy exchange occurred with the interstellar medium. The subsequent discovery of the lengthening of the pulsar periods engendered confidence in Gold's concept.

A spinning neutron star gradually slowing down over periods of thousands of years thus seems an attractive idea. But why should an object of this kind release radio energy over a wide frequency range— and why should it do so only in 'flashes' at the rotation period of the star? No one yet has a satisfactory answer to these questions, although many models have been proposed. For example, Gold's† suggestion is that neutron stars possess extremely high magnetic fields, of the order of 10^{12} gauss, at their surface. Under these circumstances any ionized gas which escapes from the star would move along the magnetic lines of force and be whirled along at the angular velocity of the star. As it moves away from the surface, so the angular velocity of the ionized gas increases until the point is reached where the tangential velocity approaches the velocity of light. (For the Crab pulsar this distance is estimated to be about 1700 km.) Gold then suggests that the plasma will radiate by the synchrotron mechanism, and since the radiation is then beamed in the forward direction, a distant observer would see this as a flashing 'lighthouse', according to the general scheme shown in Fig. 42.

A number of difficulties with this model led Graham Smith‡ to propose that the beaming was the result of a purely geometrical process,

* T. Gold, *Nature*, **218**, 731, 1968.

† T. Gold, *Nature*, **221**, 25, 1969.

‡ F. G. Smith, *Mon. Not. R. astr. Soc.*, **149**, 1, 1970; *Mon. Not. R. astr. Soc.*, **154**, 5P, 1971.

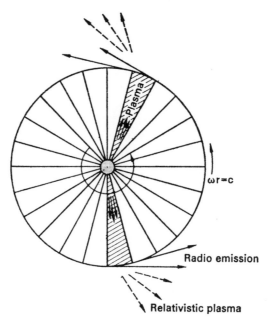

Plasma

$\omega r = c$

Radio emission

Relativistic plasma

42. Gold's idea for the production of pulses of radio energy from a rotating neutron star. The circle is where the peripheral speed ωr of the co-rotating magnetosphere becomes close to the velocity of light.

independent of wavelength. If a source of radiation moves at a velocity close to that of the velocity of light far out in the magnetosphere of the pulsar then the radiation will be compressed into a beam in the direction of travel, sweeping around with the rotation of the star. This process of relativistic beam compression seems capable of explaining many of the observed characteristics of the radio pulses, including the variability in the polarization observed from pulse to pulse.

For the energy requirements Gold assumes that the Crab pulsar is a neutron star revolving 30 times per second, that it is of solar mass concentrated in a radius of 7 km. The rotational energy is then 10^{49} erg. The spindown rate of one part is 2400 years, equivalent to a rotational loss of energy of 2×10^{38} erg per second. At a distance of 2 kiloparsecs the radio energy would be less than 10^{31} erg per second on the assumption that it came from an isotropic radiator. Thus the energy release appears to be entirely adequate to explain the observed radio energy. The rotational energy at the time of formation of the star would be much greater. If the rotation period was 1 millisecond, then 10^{52} erg is a probable figure. This is not dissimilar from estimates of the total energy release arising from the explosion of a supernova.

13

Flare stars

My own interest in flare stars arose in 1958 when at last it was possible to use the Mk I radio telescope for a research programme which included the angular diameter measurements described earlier in this book, and many other topics, such as the sky surveys and the studies of extragalactic nebulae. These and the other programmes were natural extensions of work which had already been pursued using small instruments. I felt that with this large steerable telescope available for the first time, some of our work should be directed towards opening up a new avenue of research, but it is difficult to put into practice such an idea without wasting time on hopeless enterprises. In another connection I have mentioned the Symposium on Radio Astronomy organized jointly by the International Scientific Radio Union and the International Astronomical Union, which was convened in Paris from 30 July to 6 August 1958. At that time the subject was small enough to include every topic of known interest to radio astronomy in a single meeting. The last session was on Mechanisms of Solar and Cosmic emission, and in this E. Schatzman presented a short paper on the possibility of observing radio emission from flare stars.* His conclusion that there were about a dozen stars which should emit observable radio waves during a flare served to direct my attention to this as the type of problem admirably suited to the new telescope. The flare stars discussed by Schatzman were red dwarfs and it is first necessary to explain why these stars may be of particular interest.

The life cycle of a star

The problem of the origin of the various types of stars observable in the Milky Way has received much attention from astronomers during this century. What are the fundamental features which govern the enormous range of masses, luminosities, and sizes of the stars? Today we believe that we have a reasonable understanding of the various processes by which stars generate their vast energies but before this appreciation of the

* E. Schatzman, *Paris Symposium on Radio Astronomy* (Stanford 1959), p. 552.

significance of thermonuclear processes was achieved it was known that the stars showed a general tendency to be arranged in a continuous and localized band, from those which had luminosities 100 times greater than the Sun to those of much lower temperatures with luminosities ten times less than the Sun. Hertzsprung in 1911 first plotted the luminosities of stars against their colour equivalents and in 1913, at a meeting of the American Association for the Advancement of Science, Russell presented a series of diagrams in which a large number of stars of known distances were plotted to show the relation between their luminosities and their spectral types, or temperatures. These diagrams became known as the Hertzsprung-Russell, or H-R, diagrams, and although considerably modified since that time, remain the foundation on which the current theories have been built.

As an example of the H-R diagram, Fig. 43 is based on a plot of 6700 stars, originally made by W. Gyllenberg of the Lund Observatory in Sweden. In this diagram the vertical axis is the absolute visual magnitude of the star (with the luminosity in terms of the Sun shown on the right hand side), while the horizontal axis is the temperature of the star (with the equivalent spectral type shown at the top). The clear band from the very hot blue giants at the left to the low temperature red dwarfs at the bottom right is evident. This is known as the main sequence. The Sun which is of spectral type G0 (temperature 6000°), and absolute magnitude +5, lies on this main sequence.

Although the main sequence is well defined it is evident that many types of stars—such as the white dwarfs or the supergiants—lie off this main sequence. For example Sirius is a normal main sequence star of spectral type A0, with a mass 2·4 times that of the Sun and 30 times greater luminosity, whereas its companion, Sirius B, is a white dwarf of spectral type A5, ten thousand times less luminous than the Sun, although about as massive as Sirius itself. Another example is the star Capella lying in the giant branch to the right of the main sequence, with nearly the same spectral type as the Sun (G), but 4·2 times more massive and 130 times more luminous. At the top right of the table the supergiants include some of the largest stars known (for example Betelgeuse) which have enormous luminosities many thousands of times greater than the Sun, but have low temperatures and spectral types appropriate to the red dwarfs at the bottom right of the main sequence with luminosities only a tenth of that of the Sun. A supergiant will have a diameter hundreds of times greater than that of the Sun, and hence although its temperature and intrinsic emission per unit area of surface is low, the luminosity will be very high. Thus, modern theories of star formation must seek to explain not only why the majority of stars lie on the well

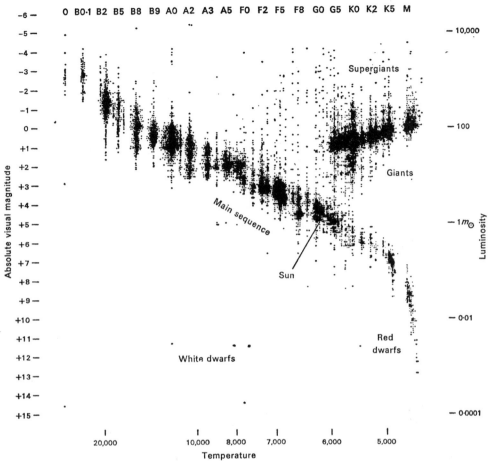

43. The H-R (Hertzsprung-Russell) diagram based on observations of 6700 stars. The ordinates are absolute visual magnitude with the luminosity in terms of the sun, as unity, on the right. The abscissae are spectral class (top) and surface temperature of the star (bottom). From the diagram originally drawn by W. Gyllenberg of Lund Observatory.

defined main sequence, but also why there are such striking exceptions lying to the left and right of this sequence.

It appears that a star lying on the main sequence is in a stable condition where the temperatures and pressures in the deep interior are such that thermonuclear processes converting hydrogen to helium provide the immense energy necessary to arrest the gravitational contraction of the star so that its external characteristics remain unchanged for long periods of time. For example, in the interior of the Sun where the temperature is about 20 million degrees and the pressure several thousand million atmospheres, the thermonuclear transformation of hydrogen to helium is converting 564 million tons of hydrogen into 560 million tons of helium every second. The 4 million tons of solar matter converted to energy every second provides the Sun with an energy output of about 4×10^{23} kW. However, since the mass of the sun is about 10^{27} tons it still has a life expectancy of several thousand million years.

Eventually, the amount of hydrogen in the core of the star will diminish and it will begin to contract in an attempt to establish a new balance. The temperature of the contracting core then rises to such an extent that other thermonuclear transformations will be initiated in the shell of gas surrounding the core. The star will then expand and move away from the main sequence to the right—to the region of the giants in Fig. 43. As the star expands its luminosity will not change greatly, and so the surface temperature diminishes. Finally, all possibilities of further energy production will be exhausted and the star will move to the left of the H-R diagram, cross the main sequence and eventually collapse into the region of the white dwarfs with radii one hundredth to one thousandth of that of the Sun and with enormous densities of 10^5 to 10^8 g cm^{-3}. However, if the star is much more massive than the Sun—at least one and half times greater mass according to present views—it is believed that the gravitational forces during the collapse are so great that the final state will be that of a neutron star. Neutron stars, already mentioned in connection with the pulsars, will have densities even greater than the white dwarfs, in the region of 10^{12} to 10^{14} g cm^{-3}. Whereas the radii of the white dwarfs correspond to that of planets, the radius of a neutron star will be only 10 to 100 km. The transition to the white dwarf state may be a reasonably quiet process, but the transition to a neutron star probably involves instabilities which lead to cataclysmic events—the supernovae.

These are believed to be the kind of processes which take place in stars lying on the main sequence. But what about the earlier history of the star? How does a star arrive on the main sequence? Originally it is presumed that the stars begin as condensations in the clouds of primeval hydrogen gas. Once a globule is formed by some instability in the cloud the mass of gas will contract under its own self gravitation, assisted by the radiation pressure of surrounding starlight. As the contraction continues, more and more of the gravitational energy of the gas is liberated and the globule will eventually become luminous. As contraction continues the temperature and pressure in the interior eventually rises to such an extent that thermonuclear processes are initiated. Probably the first thermonuclear reactions involve protons and deuterium, then when the deuterium is exhausted further contraction will take place, the temperature will rise to a few million degrees and other light elements will be involved. Finally, when the temperature and pressure have risen sufficiently thermonuclear processes involving the hydrogen-helium transformation will be initiated and this is the condition of the star on the main sequence.

The red dwarf stars

We are concerned here with the most insignificant stars—those red dwarfs lying on the bottom right of the main sequence in Fig. 43. Why have these red dwarfs reached the main sequence with such low luminosities, low surface temperatures and masses compared with the Sun? We believe that the Japanese astronomer C. Hayashi* gave the correct explanation in 1961. Previously the calculations relating to the early gravitational contraction of stars towards the main sequence described above had been made on the assumption that the equilibrium was wholly controlled by radiation. A typical evolutionary track of a star towards the main sequence, calculated on this basis, is shown as CPD in Fig. 44.

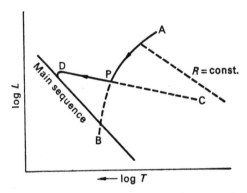

44. The evolutionary tracks of a stellar mass contracting towards the main sequence. CPD is for a contracting gas in radiative equilibrium. The Hayashi convective track is shown at APB.

The fundamental contribution of Hayashi was to show that under certain conditions convection, rather than radiation, was a critical consideration governing the equilibrium of the contracting mass of gas. He calculated that the track of a condensing mass of gas with a convection

* During my term as President of the Royal Astronomical Society it was my pleasure to announce at the Sesquicentenary anniversary meeting of the Society on 13 February 1970 that the Eddington Medal of the Society had been awarded to Professor Cushiro Hayashi of the University of Kyoto.

zone would be along APB in Fig. 44. If condensation begins in the forbidden zone to the right of APB, then the star will adjust its internal structure in a short time to move along a track with constant radius until it reaches APB. The star then moves along the track AP until it reaches P. It then moves along PD in radiative equilibrium to the main sequence. Regions to the left of AP but above PD correspond to solutions which are centrally condensed. Hence a star with uniform chemical composition and nearly uniform energy source cannot exist for a long timescale apart from the tracks AP and PD.

The evolutionary tracks and ages for stars with different masses in gravitational contraction as calculated by Hayashi* are shown in Fig. 45. The times t_1 and t_2 are the ages in years at the turning point P and

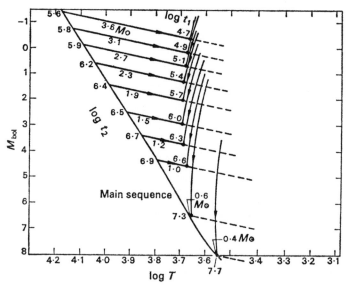

45. The evolutionary tracks towards the main sequence for stars of different masses according to Hayashi's calculations. $M_\odot$ is the mass in terms of the Sun's mass as unity. t_1 is the time in years to the turning point on the radiative track (that is the time along AP in Fig. 44). t_2 is the age of the star when it reaches the main sequence.

on the main sequence respectively. The ages of the red dwarf stars are greatly reduced compared with the earlier estimates because of their nearly vertical evolutionary track on the H-R diagram. The radius decreases rapidly so that the angular momentum of the contracting star has to be lost in a relatively short timescale. We believe that the red dwarfs which exhibit flare activity are contracting stars on the nearly vertical

* C. Hayashi, *Publ. astr. Soc. Japan.* **13**, 450, 1961.

lines of convective equilibrium in the H-R diagram. These red dwarf flare stars are believed to be contracting stars of such low mass (that is 0·4 solar masses or less) that for the whole of their luminous life time they lie on a convective track, never attaining the stage of hydrogen burning and radiative equilibrium, but passing to a degenerate black dwarf condition. During the contracting phase, the energy release per unit area of surface remains sensibly constant. Since the radius decreases rapidly, the angular momentum must be lost over a short time. Although these qualitative conditions of the red dwarf flare stars appear to be well defined, it cannot be said that there is any general understanding about the actual processes governing the flare activity or the release of the large amounts of energy which are observed during the flare phase, and it was hoped that success in observing radio emission from the flares might give a new insight into the conditions in the atmospheres of these stars.

In Fig. 45 the lowest mass where the convective track intersects the main sequence without reaching radiative equilibrium (and therefore the hydrogen burning phase) is shown as 0·4 times the mass of the Sun ($0·4 \, M_\odot$). The timescale for condensation to this point is given as about 50 million years. For even smaller masses which are applicable to some of the well known red dwarf flare stars the timescales continue to increase (for a mass $0·07 \, M_\odot$ the time for condensation is calculated to be about 500 million years). However, these timescales are about 100 times less than those which would apply to a star of the same mass contracting along a track of radiative equilibrium. The computed timescales are small compared with the age of the galaxy and hence the number of stars which have evolved to the degenerate black dwarf stage may be very large.

The flare phenomenon

These are the stars to which Schatzman was referring in his address to the 1958 Symposium on Radio Astronomy in Paris. Although we did not, at that time, understand the evolutionary track of the stars as described above, they had been recognized observationally ten years earlier. In October of 1948 Luyten* discovered the binary star L 726–8AB (UV Ceti). He measured the parallax as 0·561″ and suggested that it was the second nearest star to Earth in space. Actually the parallax was later found to be 0·375″, but the important point is that on 7 December 1948 he observed a variation of 2 magnitudes in the brightness of the star. In the meantime Joy and Humason† had studied this

---

* W. J. Luyten, *Astrophys. J.*, **109**, 532, 1949.
† A. H. Joy and M. L. Humason, *Publ. astr. Soc. Pacific*, **61**, 133, 1949.

star with the 100 inch telescope at Luyten's suggestion and had observed bright helium lines and a strengthening of the ultra-violet continuum.

In 1950 the International Astronomical Union recognized the UV Ceti class as a new type of variable. At that time the known examples of flare stars were all M type red dwarfs in the solar neighbourhood, characterized by the occurrence of sudden but irregular and short-lived increases in brightness. Increases of half a magnitude occur every few hours, of one or two magnitudes every few days, and there are records of rare occurrences of increases of 5 or 6 magnitudes on UV Ceti. Fig. 46 shows a typical flare on UV Ceti recorded photoelectrically by P. F. Chugainov in the Crimea.

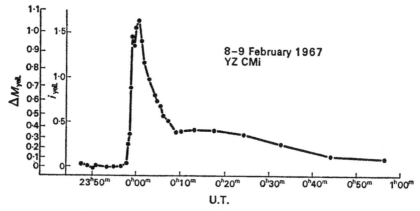

46. A typical record of a flare on a red dwarf star, showing the almost instantaneous rise to maximum brightness and the slow decay to normal. This is a photoelectric recording of a flare of about 1 mag. on the star YZ CMi (YZ Canis Minoris: Ross 882) made by P. F. Chugainov at the Crimean Astrophysical Observatory during the night of 8–9 February 1967. The horizontal scale is the time and the vertical scale shows the magnitude change ΔM above normal brightness, and the photoelectric current using a yellow filter.

There are at present some 25 stars of the UV Ceti type on which flares have been observed. With one or two exceptions they are dwarfs with dMe* type of spectra. The temperatures of the red dwarfs are below 3000 K and their main spectral peculiarities are the strong titanium oxide absorption bands. In the dMe, UV Ceti type, this characteristic spectrum is overlaid with the emission lines of hydrogen, ionized calcium, and sometimes weak helium lines are present. Generally speaking, during the flare phase the predominant effect is a strengthening of the

* A dwarf M type star with emission lines in the spectrum.

ultra-violet continuum and hydrogen lines, and the appearance of ionized helium and iron lines.

Nineteen of the 25 dMe flare stars so far observed are binaries. Of these, fourteen are visual binaries in which the flares occur on the weaker component typically situated about 1000 stellar radii away from the principal component. The masses of all the stars are small compared with the Sun. The total mass of the UV Ceti binary is about $0 \cdot 08 \, M_\odot$ and the components do not differ greatly in mass (of the order of $0 \cdot 040 \, M_\odot$ and $0 \cdot 046 \, M_\odot$). For the typical star UV Ceti, values from $15 \cdot 6$ to $16 \cdot 1$ have been given for the absolute visual magnitude M_v, a recent estimate of M_v for the fainter flaring component is $15 \cdot 88$. The orbital period of UV Ceti is about 200 years. More than a quarter of a revolution has been completed since its discovery and the separation of the components is now increasing.

Although so few flare stars of this type have been observed they must be common objects in the galaxy. Flares have been observed on about one quarter of the identified dMe stars. These emission line stars comprise about 5 per cent of all known dM objects. Because of their weakness, the few hundred known red dwarfs lie within 20 pc of the Sun, but they are believed to comprise about 80 per cent of the galactic population. Thus the few flare stars now under discussion may well be typical of the most prevalent type of stellar variation in the galaxy.

Radio observations of the flares

The exceedingly sharp rise to maximum and the slow decay to normal brightness of the typical flare shown in Fig. 46 clearly invites comparison with solar flares. However, any such straightforward comparison at radio wavelengths proves most discouraging. If we take a distance of 3×10^5 astronomical units for the nearest star, then as far as radio waves received on Earth are concerned, we have a dilution factor of nearly 10^{11} in the intensity. Outbursts on the Sun give an intensity on Earth of 10^{-19} to 10^{-20} W m^{-2} Hz^{-1} in the metre waveband. If these occurred on the nearest star, the intensity of 10^{-30} to 10^{-31} W m^{-2} Hz^{-1} would be unobservable on Earth as short-lived occurrences. There have been a few occasions since radio observations began where bursts of about 10^{-15} W m^{-2} Hz^{-1} in the metre waveband have been observed on the Sun. At the distance of the nearest star this would give 10^{-26} W m^{-2} Hz^{-1}. Such a signal of 1 flux unit is, of course, readily detectable by modern radio telescopes, but this is scarcely a practical proposition if the duration is a minute or so, occurring only occasionally every few years.

These considerations would have been totally discouraging, as far as any programme of work with the radio telescope was concerned, had it not been for the very large amount of energy which the flare stars seemed to be emitting during the flare phase. In this case general considerations are far more encouraging. The continuum emission from a typical solar flare is too weak to be seen against the photospheric background and only large flares can be seen in white light—as an order of magnitude we may take this as implying that, except for flares of great violence, a solar flare shows less than about 1 per cent of the photospheric continuum. Taking a typical flare area of 10^{18} cm^2 with a duration of 1000 sec, the total energy would then be about 10^{22}–10^{23} joules. Occasionally very great flares (such as the one on 26 July 1946) show a 10 per cent increase in the continuum and then the total energy rises to 10^{23}–10^{24} joules.

In the two magnitude flare observed on UV Ceti by Luyten the flare energy was estimated to be 4×10^{24} joules. Observations of flares on other red dwarf stars gave similar total energies. Thus, it seemed that the energies involved in the relatively frequent flares of about 2 magnitudes on the red dwarf stars were in excess of the energies of the most violent solar flares, and were 100 to 1000 times in excess of the more frequent solar flares.

It was on the basis of these estimates that I began observing flare stars in the metre waveband with the radio telescope at Jodrell Bank, the expectation being that it might be possible to observe increases in intensity of at least a flux unit every 20 hours or so of observing time. The observations of UV Ceti began on 28 September 1958. There was little to guide one on the choice of receiver frequency. However, on the assumption that there was some similarity with solar flares, the metre waveband was indicated. Furthermore, since long periods with the telescope in automatic motion were needed, a frequency giving a beamwidth of not less than a degree was desirable, so that minor following errors introduced by the analogue computer would not be detrimental. In any case, at that time radio frequency equipment with good noise factors on frequencies higher than a few hundred megahertz was not available. The other problem was to find a frequency which was free from radio interference. By good fortune Hazard had, at that time, a system working on 240 MHz for his lunar occultation measurements. Although not a band reserved for radio astronomy, it was secured for military purposes, and at that time was almost completely free at Jodrell Bank. In principle, therefore, the procedure was simply to drive the telescope on UV Ceti and record the output on the pen recorder. This was how my observations began and immediate encouragement was derived from a burst of a few flux units of short duration, occurring between 01 and 02 UT on 29 September 1958.

Unfortunately, it soon became clear that the type of low-intensity short-duration signal for which I was searching was only too readily produced from many terrestrial sources. After some experience I placed another offset aerial on the telescope and recorded through two channels, one on the star and the other on an adjacent comparison area of sky. In this manner 474 hours of observation were made on 5 flare stars between 28 September 1958 and 14 April 1960*. When all doubtful cases had been excluded, there were 13 events compatible with bursts of radio emission from the flare star, appearing on the star channel but not on the comparison channel. On UV Ceti, which was used for nearly half the observations, there was one probable event in 35 hours of observation, and this was close to the rate of occurrence of visual flares of a few magnitudes (1 in 35 hours).

It had been realized that results of this nature needed simultaneous visual or photographic records of the star, and after the initial records had been obtained, a collaborative programme was arranged with Dr. Dewhirst at the Cambridge Observatories. However, poor sky conditions led to such a low yield of mutual photographic and radio coverage that it was not considered profitable to pursue the programme. It was therefore fortunate that I showed these records to Professor Fred L. Whipple when he visited Jodrell Bank in the summer of 1960, in order to elicit his interest in a collaborative programme. He immediately suggested that the Baker–Nunn cameras of the Smithsonian's Satellite Tracking network should be used in a collaborative programme.

There were five stations in this network conveniently distributed around the longitude of Jodrell Bank—in Iran, South Africa, Spain, Curaçao, and the Argentine. The cameras were 20 inch aperture f/1. Initially, the exposure was for less than 15 seconds on 1 frame every 3 minutes, subsequently reduced to 1 frame in 2 minutes. The magnitude of the flare star was measured by a visual comparison method with a standard deviation of about 0.1^m for good quality film and 0.2^m for poor quality. A magnitude change of less than 0.3^m was not regarded as significant. The details of the observing programme and the reduction of the films were in the hands of Mr. Leonard Solomon, one of Professor Whipple's staff at the Smithsonian Astrophysical Observatory.

This collaborative programme began on 28 September 1960—using the 240 MHz equipment—precisely two years after my initial radio observations. A year later, after 6 sessions, each lasting for about 14 nights of the no-Moon period, we had accumulated 727 hours with the radio telescope. Just under a third had been covered by the cameras and there were 166 hours of overlap where both radio and camera records

* The details have been published in *The Observatory*, **84**, 191, 1964.

were of good quality. Of this, 125 hours was on UV Ceti and the remainder on Ross 882 and EV Lac. During this whole period there was no bright flare. There were 23 minor flares of less than 1 magnitude and since there was no obvious deflection on the radio charts at these epochs, they were integrated by superimposing the radio records aligned on the epoch of the peak of the photographic event.* The result of the integration is illustrated in Fig. 47 which shows the increase in radio emission over the interval −2 minutes to +8 minutes about the epoch of the flare maximum. Comparison with a similar integration of the charts where no flare occurred shows that the chance that the increase is spurious is less than 1 in 10^8.

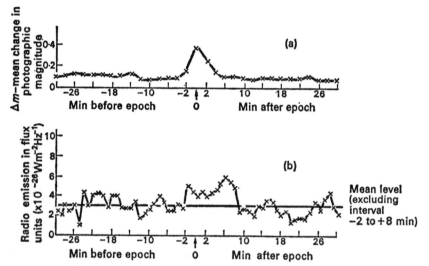

47. The first conclusive association of radio emission with flares on the red dwarf stars. These records are the superimposition of 23 cases of minor flares (a) as observed photographically with the Baker-Nunn Satellite tracking cameras of the Smithsonian Astrophysical Observatory and (b) as observed with the radio telescope at Jodrell Bank. The observations were made in 1960–1. In both (a) and (b) the horizontal scale is time with the epoch of maximum of the photographic flares taken as the zero point. The ordinates in (a) are change in photographic magnitude and (b) change in flux units on 240 MHz.

The collaborative programme using the Smithsonian cameras continued thereafter on a regular basis, but it was clear that the programme was not efficient in terms of radio telescope time. Even with the five cameras, considerable time was lost because of poor sky conditions. There were also many gaps when the tracking of satellites had priority. By March 1963 we had completed 14 separate runs, mostly on UV Ceti and Ross 882 (YZ CMi). The radio telescope had been used in obser-

* B. Lovell, F. L. Whipple, and L. H. Solomon, *Nature*, **198**, 228, 1963.

vations for 1444 hours, but the good overlap with the camera records was only 307 hours. Relatively little of this time was lost because of radio telescope problems, since the 307 hours of camera overlap was from a total of only 398 hours of camera observing. Thus a factor of between 4 and 5 remained to be gained in effective observing time and in view of the rarity of the events under observation an improvement in efficiency was much to be desired.

It was my good fortune at that time to visit the Soviet Union during the summer of 1963 as the guest of the Academy of Sciences (see Chapters 6 and 14). I found many of the Soviet astronomers were greatly interested in the problem of the flare stars. At Burakan, Academician V. A. Ambartsumian was working on the problem with V. Oskanjan from Belgrade. At the Crimean Astrophysical Observatory, P. F. Chugainov had excellent photoelectric recording equipment (an EMI 6256 B photomultiplier with yellow and blue filters) attached to a 64 cm meniscus telescope. The Academy readily agreed to a joint observing programme, and collaborative arrangements were instantly made with the Crimean observatory, with the Abastumani Observatory in Georgia (using a 40 cm Schmidt camera of 62·5 cm focal length, with orthochromatic film and filters), and with the Odessa observatory where telescopic observations were made at $\frac{1}{2}$ minute intervals.

By the autumn of 1963 this was tied in simultaneously with the Smithsonian network and the programme of observations instantly achieved an entirely new outlook. In the first attempt from 18 September to 3 October 1963 I observed UV Ceti for 129 hours with the radio telescope and the effective overlap with the Smithsonian and Crimea+ Odessa+Abastumani records was 102 hours. These collaborative programmes were then carried out regularly until December 1969 when they had to be suspended temporarily because of the engineering changes which were beginning on the Mk I telescope. Up to that time 36 separate observing programmes had been run involving over 4000 hours of observing time with the radio telescope. It has only been possible to analyse and publish a part of the information which has accumulated.* There are many cases of radio/optical correlations of individual flares. The record of an unusually long duration flare is shown in Fig. 48 in which the flare (on YZ CMi) lasts for several hours. Good fortune attended the international collaborations on that night. Late in the afternoon of Saturday 18 January 1969 I calibrated the apparatus on the standard source 3C 33 (a quasar) which I normally used, and

* For more detailed accounts see B. Lovell, *The Observatory*, **84**, 191, 1964. *Q.J.R. astr. Soc.*, **12**, 98, 1971.

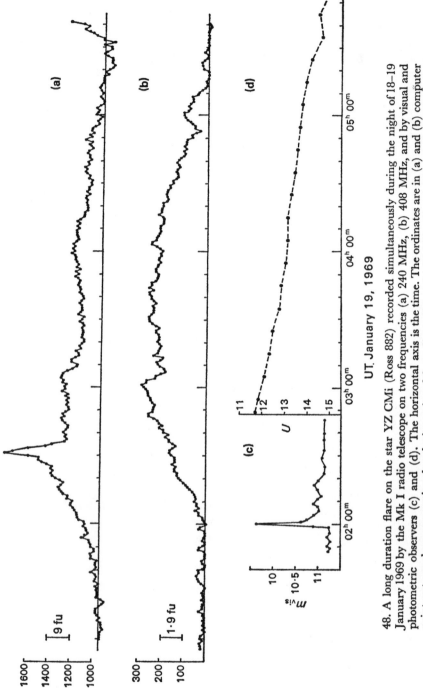

48. A long duration flare on the star YZ CMi (Ross 882) recorded simultaneously during the night of 18–19 January 1969 by the Mk I radio telescope on two frequencies (a) 240 MHz, (b) 408 MHz, and by visual and photometric observers (c) and (d). The horizontal axis is the time. The ordinates are in (a) and (b) computer print-out numbers proportional to the intensity of the radio emission (in (a) 200 units = 9 flux units; in (b) 100 units = 1·9 flux units; 1 flux unit = 10^{-26} W m^{-2} Hz^{-1}). In (c) the ordinate is the increase in visual magnitude and in (d) U-band photometric record—the conversion of the peak of the flare in (c) to this scale would be at U = 8·5 approximately.

then asked the controller to drive the telescope to the flare star Ross 882 (YZ CMi). He reached the correct position and began following the star at 5.49 p.m. when it was just over 1 degree above the horizon. After waiting for everything to settle down I had to leave. The controller was asked to maintain the telescope on the star and I said that I would call in later. When I did so it was nearly midnight and to my dismay I found that the pen recorder had drifted off scale.

Already at that time, unknown to me, Chugainov in the Crimea had observed one flare on the star and my apparatus had recorded it also at 6.15 p.m. before drifting off the chart. Having reset the apparatus I retired from the scene once more. At 2 a.m. Dr. Andrews phoned from Armagh to say that a great flare was in process. He and his colleagues observed for an hour and believing the event to be over, ceased observations. Within minutes of the cessation of their observations, W. E. Kunkel began his observations at Cerro Tololo. My own records were extremely good, primarily because of the lack of traffic and other disturbing factors during the early hours of a Sunday morning. Kunkel was able to continue his optical observations until 6.15 a.m., when the flare had finally decayed below the detection threshold. At about the same time the star set below the Jodrell horizon. Through these various fortuitous circumstances we obtained a unique record of radio and optical measurements of the flare throughout the whole night shown in Fig. 48.

Flare stars and the galactic background radio emission

There have been periods in the earlier development of radio astronomy when it seemed that stars or star-like bodies were responsible for the emission of the radio waves received on Earth. The initial belief that the ordinary stars were sufficiently strong sources of emission was soon shown to be erroneous when the early workers failed to detect radio emission from them. However, the discovery of the localized sources, before the extragalactic nature of many of them had been established, led to a revival of the star concept. In the 1949–51 period comprehensive theories were developed on the basis that the radio waves were emitted from a distribution of stars or star-like bodies. Then the evidence began to accumulate that many of the radio objects (at that time called radio stars), were extragalactic. At the same time the Soviet astronomer I. Shklovsky developed his theory that the background emission was a mixture of the two processes of synchrotron and free-free emission in the ionized hydrogen gas. These developments led to the final decay of the radio star concept as an explanation of the continuum emission from the galaxy. Notwithstanding the apparently conclusive nature of this evidence, the discovery of the radio emission from the red dwarfs again raises the question of a possible star-like contributor to the generally distributed galactic emission—particularly since the red dwarfs probably represent a high proportion of the stars in the galaxy.

The appropriate comparison is with the observed brightness temperature T of the radio emission from the galaxy, which is a marked function of the wavelength, and the latitude and longitude. Most of the flare star data have been obtained on a frequency of 240 MHz. The observed value of T at this frequency varies from 50 K in the direction of the anticentre to several hundred degrees along the disc towards the centre, but it is believed that 14 K may be contributed by extragalactic sources, and 3 K by the background emission (Chapter 9). In this case the galactic contribution is 33 K. Detailed calculations of the amount of this arising from the integrated flare star emission depend on the assumptions which are made about the distribution of the red dwarfs. If they are localized to the solar neighbourhood, extending say to 1 kpc, then the contribution is small—only 0·13 K. On the other hand if the distribution observed in the solar neighbourhood is assumed to extend throughout the galaxy then the contribution becomes more significant and rises to 2·6 K for a distribution out to 20 kpc.

The energy of the outbursts

The early investigators of the flare stars found that the total energy of the flares was surprisingly large. On the Sun only extremely large flares are seen in white light against the photospheric continuum. A typical large flare lasting for 1000 s and occupying 10^{18} cm^2 would have an energy of not more than one per cent of the energy emitted by the photosphere. The photospheric emittance is $6·35 \times 10^{10}$ erg cm^{-2} s^{-1} and hence the total energy involved in large solar flares is of the order of 10^{23} joules or less. Over a similar period of 1000 s the total energy of the photospheric emission would be about 10^{29} joules. Thus, in the familiar case of the Sun, we are dealing with eruptive processes which, energetically, are equivalent to only about a millionth of the normal energy emission.

In the case of the flare stars we find an entirely different situation. In the flare which we observed on YZ CMi on 19 January 1969, Kunkel's measurements[*] gave the total energy in the optical spectrum to be about 10^{27} joules at an average rate of 10^{23} joules per second. But the total energy radiated by that star in its quiescent state is only 10^{23} joules per second. Thus, in this case there was a catastrophic energy release, at the rate of the normal energy release from the star, which lasted for several hours after which the star returned to its normal brightness. Indeed, in these red dwarf flares this is a typical condition, where the catastrophic

[*] W. E. Kunkel, *Nature*, **222**, 1129, 1969.

energy release is at the same rate as the normal rate of energy release by the entire star, although the duration is usually less than the example given above. Of course it is precisely these features of the red dwarf stars which justified the optimism mentioned on page 176 that it might be possible to observe them in the flare phase with the radio telescope.

The conditions in the atmosphere of the stars

The ability to study the radio emission from the Sun quickly led to a great clarification of the complex conditions in the solar atmosphere and corona. It may be hoped that, in due course, the ability to study the radio emissions from the red dwarf stars will lead to a similar increase in our knowledge of the conditions in the atmospheres of such small condensing masses of gas. Earlier in this chapter the current beliefs about the basic nature of these contracting globules have been mentioned, and in a general sense, at least, it appears that the rapid decrease in the radius of the contracting mass of gas would lead to conditions where large amounts of energy would be released.

At this moment no one would claim to be able to give a satisfactory explanation of the processes by which the red dwarfs release such great amounts of energy so abruptly. The study of the radio flares lead us to believe that a shockwave travels out from the explosive centre of the flare—although the reason for the occurrence of such explosions in the star are uncertain. It has been fortunate that Professor Kahn and his colleagues in the Department of Astronomy of the University of Manchester have been interested in my own observations of the flares and a reasonably coherent theory of the mechanism by which the radio flares are produced seems to be emerging.* Kahn believes that the explosion causes a shock wave to travel out through the star's atmosphere at supersonic speed (about Mach 2), and that radio emission is produced when the shock reaches the level in the star's atmosphere where the electron density is appropriate for excitation at the plasma frequency. These conditions are that

$$2\pi\nu = \left[\frac{4\pi Ne^2}{m}\right]^{\frac{1}{2}}$$

where ν is the radio frequency excited, N is the electron density, e and m are the charge and mass of the electron. For radio emission at 240 MHz this determines that the excitation takes place when the shock reaches the level in the stellar atmosphere at which $N = 7 \times 10^8$ electrons cm^{-3}. The 408 MHz radiation will be produced at a level where $N = 2 \times 10^9$ electrons cm^{-3}.

* F. D. Kahn, *Nature*, **222**, 1130, 1969.

In the case of the flare on YZ CMi shown in Fig. 48 Kahn's calculations give numerical values for the temperature and level in the stellar atmosphere at which these radio emissions occur. For this star the radius of the photosphere is believed to be one quarter of that of the solar radius, that is $R_\odot/4 = 1\cdot75 \times 10^{10}$ cm. The 408 MHz emission is then found to occur at a radius in the stellar atmosphere of $2\cdot5 \times 10^{10}$ cm, and the 240 MHz emission at a radius of $3\cdot5 \times 10^{10}$ cm. The temperature at these radii is calculated to be about $2\cdot8 \times 10^6$ K.

In fact, there are other general considerations relating to the gravitational stability of the star which show that the temperature in the atmosphere cannot be significantly greater than this value. For this star $R \simeq 0\cdot25\ R_\odot$ and $M \simeq 0\cdot3\ M_\odot$, and hence the escape velocity must be nearly the same as that for the Sun, that is

$$V_\odot = \sqrt{\frac{2GM_\odot}{R_\odot}} = 6\cdot17 \times 10^7 \text{ cm s}^{-1}$$

Thus the effective temperature in the region where the flare emission occurs, in YZ CMi, cannot be greater than about 10^5 or 10^6 K.

With these conditions at a level of about 3×10^{10} cm in the star how much of the atmosphere will be blown away into space as a stellar wind? Kahn calculates the normal rate of loss of mass for the whole star carried away by the stellar wind, as $3 \times 10^{-12}\ M_\odot$ per year, which is satisfactory since it is too low to have any marked effect on the period of evolution of the star. From an analysis of the properties of the optical flares the Soviet Astronomer R. E. Gershberg has concluded[*] that the rate of loss of mass is less than $10^{-11}\ M_\odot$ per year, which is in satisfactory agreement with Kahn's calculation. Although these loss rates are small for the individual stars, nevertheless collectively, because of the very large number of red dwarfs, the total contribution from the red dwarfs over the whole galaxy makes a predominant contribution of stellar material to the interstellar medium, being 10 to 100 times greater than that contributed by novae and other eruptive processes.[†]

These arguments strengthen the view that the star flares and the solar flares are essentially similar phenomena, being primarily localized events in different levels of the chromospheric regions, with markedly similar physical manifestations in the optical and radio domains. The energy source for the solar flare is the magnetic field, and the activity is determined by magneto-hydrodynamic phenomena arising from convective

* R. E. Gershberg, *Problems of Stellar Evolution and Variable Stars* (Nauka, Moscow 1968), p. 50.

† A. A. Boyarchuk, *Interstellar Gas Dynamics*, I.A.U. Symposium No. 39 (Reidel 1970), p. 281.

motions. Since the red dwarf stars are in a completely convective state, it is natural to expect strong magneto-hydrodynamic effects and—in view of their small mass compared with that of the Sun—far more pronounced flare effects.

It is hard to predict at this stage whether the radio observations of the flare stars will eventually stimulate a new attack on stellar phenomena, as happened in the case of the Sun. The problem is simply that the star flare phenomena are difficult to observe even with the largest and most sensitive radio telescopes available today. The world-wide technical trend has been to smaller, more accurate telescopes, for use at very short wavelengths, and because of the spectral characteristics of the radio emission these would not appear to be adapted to such studies. For observation of the eruptive phenomena, a steerable telescope larger than the Mk I is required, working in the frequency range of a few hundred megahertz. There are unlikely to be many such telescopes built in the world in this century. However, the large aperture synthesis arrays such as the instrument at Westerbork in Holland and the new system in Cambridge, might well reach a level of sensitivity such that the radio emission from the quiescent atmosphere of the nearby stars could be observed.*

* In fact this is precisely what happened in 1972. In the centimetre waveband variable radio emission from β-Persei (Algol) was found by the observers at N.R.A.O. and in the autumn of 1972 Ryle used the radio emission from this star to calibrate the new Cambridge aperture synthesis telescope. At Bonn with the new 100 m radio telescope using a wavelength of 2·8 cm, radio emission from a red giant (Betelgeuse) was found.

14

Venus—and collaboration with the Soviet Union

I have referred already to my visit to the Soviet Union in 1963. A fruitful result of that visit was the close collaboration on the flare star observations, described in Chapter 13, which still continues. A second stimulating, but less successful attempt at collaboration concerned the long baseline interferometer. When describing these interchanges in Chapter 6, I mentioned a third collaborative arrangement—a bistatic radar experiment on the planet Venus. It is remarkable that this enterprise succeeded in spite of great organizational and technical difficulties.

There are two separate developments which placed me in the surprising position of proposing this experiment. On 12 February 1961 the Soviet Union launched a space probe towards the planet Venus. Nowadays little notice is taken of the launching of unmanned lunar or planetary probes, but in 1961 even the fact that the Soviets could attempt such an enterprise was surprising. They had already launched three successful lunar probes. The U.S.A. had tried for the Moon with four Pioneers. Only one travelled significantly into space—Pioneer 4, which missed the Moon by 60 000 km in March 1959. Pioneer 5 was destined for Venus, but in the event was launched into space with no specific target in March 1960.* Both the scientific and general interest in the Soviet Venus 1 was, therefore, intense. After some days we received from Moscow detailed information about the transmission frequencies, and trajectory of the probe. The Soviet lunar probes had worked on a frequency of 180 MHz. The transmission frequency of the Venus probe was 922 MHz. By the time we had assembled the new equipment, messages from Moscow left no doubt that the Soviets had lost contact. By early March there was detailed collaboration. The Soviets were sending us the times during which they interrogated the probe in the hope that with our great sensitivity we could locate the signals. From the com-

* For a more detailed account of the association of Jodrell Bank with these early lunar and space probes see *The Story of Jodrell Bank*.

puted trajectory it was estimated that the probe would be near Venus between 19–21 May. On 17 May we recorded transmissions which might have come from the probe. On 25 May Professor Alla Massevich telephoned me from the Soviet Union complaining that the recording which we sent had not yet been received at the Academy. A few days later another call informed me that the Academy wished her and Dr. Khodarev, the designer of the probe, to visit Jodrell, to make an attempt themselves to locate the probe. In June they came and there followed a fascinating liaison by telephone between our small laboratory and their own control centre 'somewhere in the Soviet Union'.

The Soviet scientists were very happy with this practical collaboration and there seems little doubt that it was this extension of our collaboration which led to the invitation for me to go to the Soviet Union two years later. The invitation from Academician Keldysh mentioned various visits to observatories and lectures. I arrived in Moscow on 25 June and was rushed from the airport to the enormous press conference convened for the astronauts Bykorsky and Valentina Tereschova who had recently returned to Earth after their space flight in Vostok 5 and 6 respectively. The next day when I met Academician Keldysh and his colleagues at the Academy, the unbelievable news was conveyed to me that I would be taken to their deep-space tracking station—hitherto so thoroughly screened from Western eyes.

Many years later some details of the position of this establishment and photographs of the tracking telescopes appeared in the American press. But for me at that time, it was truly a journey into the unknown realms both of Soviet territory and science. First I was taken to the Observatory in the Crimea where the young astronomers were meeting and where the discussions with Shklovsky about the structure of the quasars, mentioned in Chapter 6 occurred. Then on 1 July we set off on a long car journey through Simperofol, the health resort of Saki, along the Black Sea coast, where we cooled ourselves in the sea, and Khodarev pointed to a far-off peninsula and said his antennae were there. After many more dusty roads the huge aerials of the tracking station slowly materialized. I should have been familiar with large telescope structures but these seemed quite immense. There were three separate instruments each made up of eight 16 metre paraboloids mounted on a massive 'bedstead' framework, the whole completely steerable in azimuth and elevation. A decade has not dimmed the memory of that visit, always surrounded by 10 or 20 Russians seemingly talking simultaneously while I tried to grasp the main features of that extraordinary place.

Deep underground, adjacent to one of the telescopes, I was conducted through an enormous transmitting hall containing six 20 kW klystron transmitters. More than 100 kW of energy at 700 MHz was being fed into the telescope for the control of the space probes—so lethal to humans that it could never be used below 10 degrees elevation. It was this installation which led me to suggest a collaborative experiment during which the Soviet scientists would transmit signals to the planet Venus and we would use our telescope at Jodrell Bank to receive them after reflection from the planet—a round trip from the Crimea to Jodrell Bank of some 60 million miles. Before I finally left Moscow this, and the other 2 collaborative enterprises, had been agreed with Academician Keldysh and with Academician Kotelnikov, the director of the Institute of Radio Technics and Electronics who was responsible for the establishment in the Crimea.

The second development concerns the scientific reasons which led me to emphasize the value of the experiment at that moment. An intriguing prospect which had been opened by radar developments during the second world war was the possibility of building a radar system powerful enough to enable radar echoes to be obtained from the Moon. This was certainly one of our early ambitions at Jodrell Bank, but in the event, we were not the first to succeed. In fact, success was reported almost simultaneously early in 1946 by a Hungarian, Z. Bay, and the United States Army Signal Corps. The equipment used by Bay was unconventional in that the signals were displayed on a battery of water voltameters. Most of the voltameters received set noise only, but one also received the reflected signal from the Moon. This showed an excess of hydrogen over the others after 30 minutes. It is not clear what use could be made of such a system apart from the detection of energy reflected from the Moon after integrating for a long time. The U.S. equipment was a conventional radar on a wavelength of 2·6 metres using long pulses and a large aerial array. The Australians were the next to succeed by using the transmitter of the broadcasting station 'Radio Australia'. In the event we had to take fourth place, but as distinct from the earlier workers, when success was achieved in the autumn of 1953, we were using a steerable aerial so that the Moon could be followed continuously. Both the Americans and Australians had been puzzled by the variability of the signals reflected from the Moon. The latter had suggested that the effect was due to the movement of the Moon relative to the Earth, known as libration. The studies of the phenomena at Jodrell settled the issue. There were two kinds of fading—a rapid one with periods of seconds which was caused by the libration of the Moon, but this fading was superimposed on a much longer period fading—up to 30 minutes. We were able to show that this long period fading was an effect caused by the rotation of the plane of polarization as the radio waves traversed the Earth's ionosphere.

The Mk I radio telescope was far larger than any previous aerial used in this work and obtaining lunar echoes was a simple matter. In fact, the telescope was first tested in this way when we brought it into use to locate the carrier rocket of Sputnik I in October 1957*. Subsequently much work was done with the Mk I on lunar problems. These included the use of the fading of the lunar echoes associated with the rotation of the plane of polarization to study the variation in the total electron content of the ionosphere along the line of sight from Earth to Moon. There were also bistatic experiments with an American system—that is, transmissions from Jodrell Bank and reception in the U.S. via the Moon or vice versa. These bistatic experiments were related to the possibility of using the Moon as a passive reflector for transmission of speech and will be referred to in Chapter 15.

The ease with which lunar radio echoes could be obtained with this new Mk I telescope stimulated us to far greater fields of endeavour. In particular, the planet Venus presented an exciting prospect. Apart from the technical accomplishment of sending a radio wave into space and observing the reflected wave from the planet four minutes later there were two extremely important scientific issues. First, there was the outstanding problem of the distance of the Sun from the Earth, or the value of the solar parallax—that is, the angle subtended by the radius of the Earth at the Sun. This unit is of fundamental importance in astronomy since the Earth–Sun distance forms the baseline for the measurements of the distances of the nearer stars—and hence the distance scale for the universe. Halley's famous observations of the transit of Venus across the Sun's disc in 1716 was one of the first attempts to make measurements of the times of transit at points on the Earth's surface widely separated in latitude. In 1769 a value for the solar parallax of 8·75 seconds of arc was obtained. At the next transits in 1874 and 1882 expensive expeditions were organized both by the U.K. and the U.S.A., but the recorded times differed by 10 seconds. It seemed impossible to make a precise timing of the moment when Venus touched the Sun's disc—probably because of the atmosphere of the planet.

A more reliable method was used by Gill in 1877, who observed the relative displacement of a planet against the stars as seen from two points on the Earth's surface. Gill used various planets, first Mars, obtaining a value for the parallax of 8·78 seconds of arc, and later several minor planets to avoid the difficulties introduced by the disc of the planet. By 1900, using Eros, the value had been refined to 8·790 seconds of arc. In

* See *The Story of Jodrell Bank*, Ch. 30.

this century more subtle methods were used; for example, observations of perturbations caused by the Earth's attraction in the motion of the planets Venus or Eros, or the determination of the constant of aberration by spectroscopic measurements, at half yearly intervals, of the relative velocity between the Earth and a star.

The net result of all these attempts to determine the solar parallax was that the two most accurate methods were estimated to be correct to 1 part in 10 000, but their results differed by 1 part in 1000. That is, the fundamental unit of distance in the universe was known only to an accuracy of 0·1 per cent. The attraction of the radar method was simple to understand. If a radar pulse could be transmitted to the planet Venus and observed after reflection, the time of the journey—and hence the distance to the planet—could be measured with precision. The Earth–Sun distance could then be computed precisely by using Kepler's laws—since the periods of the planets were accurately known.

The second problem was that of the rotation rate of the planet Venus. The thick cloud enveloping the planet had frustrated all efforts to measure how fast the planet was rotating on its axis—or even in which direction it was rotating. Various theoretical estimates covered the range of possibilities from a few hours to nearly a month. In this case the radar method held the prospect of observing the changes in the scattered radio wave caused by the differential velocity of approach and recession at opposite limbs of the planet.

Although these general possibilities were readily appreciated, the technical problem appeared formidable at the time when the Mk I came into use in 1957. A simple calculation indicated the order of magnitude of the difficulty. The signal received after scattering from a distant body with given transmitter, aerial and receiver parameters is proportional to σ/R^4 where σ is the equivalent echoing area of the target and R the range. For the Moon $R = 3\cdot84\times10^5$ km, the radius $r = 1\cdot74\times10^3$ km. For the planet Venus at close approach $R = 3\cdot7\times10^7$ km, $r = 6\cdot1\times10^3$ km. If we assume that the reflection coefficient of the planet is the same as that of the Moon then, for comparison, we take $\sigma = \pi r^2$ in each case. Then the value of σ/R^4 for Venus relative to the Moon is 2×10^{-7}. In other words the overall sensitivity of a transmitter, aerial, and receiver system, which can study the Moon by radar, would have to be improved by about 5 million times to achieve similar results on Venus (and by 100 million times for Mars and 3000 million times for Jupiter on the same assumptions).

Since σ and R are set by the characteristics and distance of the planet, it is necessary to obtain such increases in sensitivity by improvements to the transmitter, aerial or receiver. It can readily be calculated that the signal to noise ratio in the receiver of a system used for planetary radar is proportional to

$$\frac{PA^2}{\lambda^2 BT} \cdot \frac{\sigma}{R^4}$$

where P is the transmitter power, A is the area of the radio telescope (assumed to be used both for transmitting and receiving), λ is the wavelength and B is the bandwidth of the receiver. σ and R refer to the effective scattering area and distance of the planet. T is the system noise temperature. At low frequencies (say to 200–300 MHz) the limit is set by the noise power received from the sky. At higher frequencies the limit is set by the noise level in the receiver.

Obviously for maximum signal to noise ratio the factor $PA^2/\lambda^2 BT$ should be as large as possible. Unfortunately, many factors interweave. In principle it is hardest and most expensive to increase this ratio by increases in transmitter power P and aerial size A; and easiest and cheapest to decrease λ, B and T. However, as λ is decreased so the surface of the paraboloid (if this is used as an aerial) has to be more accurate and it becomes difficult and more expensive to increase A. The value of T is set either by nature (at low frequencies) or by the available parametric or maser radio frequency amplifiers at higher frequencies. The bandwidth of the receiver B is the simplest to adjust, but limits in this case are set by the width of the transmitted pulse τ (in a pulsed radar). In order to accommodate the spectrum of the pulse, B must be at least $2/\tau$. A limit to the useful value of τ is set by the planet. The radar depth of Venus is 41 milliseconds. A pulse width τ of 5 milliseconds is already a somewhat coarse pulse for exploration of the planet and this sets a limit of about 400 Hz for B. If better resolution is obtained by decreasing τ, so B increases and the signal to noise ratio falls.

If a continuous wave (CW) system is used instead of a pulsed radar, then the intrinsic bandwidth of the transmissions can be very small (for Venus it would be the light-time to Venus and back, that is $2R/c$ or about 300 seconds). However, it is not possible to use the appropriate small value of B because the rotation of the planet imposes a bandwidth on the echo. From limb to limb this bandwidth is

$$4r\left(\frac{2\pi}{t}\right)\frac{f}{c}$$

where r is the radius of the planet, t the period of rotation and f the radio frequency. The optimum value of B is usually taken as some fraction of this, since only a part of the disc of the planet is likely to be effective as a scatterer.

With the optimum range of the above parameters which could be envisaged for the Mk I telescope, the calculations of the signal to noise ratio to be expected were discouraging. Fortunately there remained the strategem of integration. In a pulsed radar the improvement in signal to noise ratio is proportional to the square root of the number of pulses integrated. In the CW system the improvement is inversely proportional to the square root of the bandwidth and integration time. The ability of the Mk I telescope to follow the planet continuously enabled us to think in terms of long integration times.

Our first problem was to find a transmitter of high-power working within a frequency range for which the surface of the telescope would be efficient, and our second problem was to arrange for the power from such a transmitter to reach the aerial feed without significant loss. The usual combination of circumstances which seemed to favour us so much during those early years with the telescope conspired to help once more. To begin with I had an outstanding young man who was full of enthusiasm for using the telescope in the radar mode. He was J. V. Evans, one of our 1954 Manchester graduates. He came to Jodrell Bank as a D.S.I.R. maintenance grant student in October of that year. Before the telescope came into use he had used other equipment and had settled the problems surrounding the rapid fading of the radar echoes from the Moon. He began to make a great impression both on us, and unfortunately on our American visitors. However, before the Americans finally claimed him in the spring of 1960 he had left his mark on the new telescope—in the radar work on the Moon, the Sputniks, in a Moon communication circuit (see Chapter 15), and on the planet Venus.

Next, I discovered that my colleagues in the Physics Department had a giant klystron in which they were beginning to lose interest. This klystron worked on a frequency of 408 MHz, and was capable of delivering 100 kW in pulses of several milliseconds at a mean power level of several killowatts provided it was properly cooled. It was in use on a linear accelerator for the production of high-energy protons. A new and better accelerator of greater capacity was being installed and I eagerly seized the klystron. With all its clumsy cooling system and vast array of huge condenser banks (which I bought at scrap prices from some R.A.F. disposal) the nearest point we could get it to the aerial was in one of the tower tops of the telescope. From there the radio frequency pulses had to go through the trunnion bearing via a rotating joint, and then another 200 ft, through the best cable we could afford, to the aerial at the top of the tower.

The driver stages and associated equipment were in one of the ground level huts some distance from the telescope. The weakening of our normal electrical supply at each massive pulse was so severe that we brought up a special 60 kW mobile diesel generator to feed that hut. By the time of the close approach of Venus in September 1959 we were ready to make the attempt to detect the planet by radar. Unfortunately, the feeder systems from the transmitter to the aerial failed to handle the full power of the klystron and in the event the experiment was made with a transmitted power of only 50 kW. The pulse length was 30 milliseconds and one pulse was transmitted every second. The technique was to transmit with the telescope following the planet for a period of about 5 to 6 minutes (the time of travel of the pulse to and from the planet), and then switch off the transmitter and receive for a similar period. Since the receiver had a noise level of 4·5 db, and there were losses of 2·5 db in the cables, it was hardly surprising that no echoes were detected stronger than the noise level of the equipment. However, an integrating technique with a display on Post Office counters was used to add together the receiver noise powers corresponding to the same range intervals on successive sweeps of the time base. Eight adjacent range intervals were examined, each equal in width to the 30 ms transmitter pulse. Adjustments were made to compensate for the change in range of the planet so that any echo would remain in the same 30 ms time interval from day to day. Compensation was also made for the doppler shift in frequency of the echo, relative to the transmitted frequency.

350 transmit-receive periods, equivalent to $58\frac{3}{4}$ hours, were integrated in this manner with the result that one of the eight range intervals showed an excess count of $2\frac{1}{2}$ times the standard deviation of the counts in the 8 channels. Did this indicate that a planetary echo had been received? Dummy runs without the transmitter established the deviation throughout the 8 channels of the integrating system and it was concluded that there was an 8 per cent chance that noise alone would produce the observed $2\frac{1}{2}$ standard deviations excess in one of the eight channels. If true planetary echoes were responsible for the excess counts, then the range observed gave a value for the solar parallax of $8·8020\pm 0·0005$ seconds of arc. This in turn gave a value for the astronomical unit which was 60 000 km less than the hitherto generally accepted best value from the previous optical and dynamical measurements.

In the previous year, workers at the Lincoln Laboratories of the M.I.T. had used a smaller radio telescope with a better receiver and much more powerful transmitter. They had obtained identical results with a somewhat similar low statistical certainty. Now although the statistics of both observations were suspect, the chance that, by coincidence, the excess noise powers should appear in the same range interval in both the M.I.T. and Jodrell observations was computed to be only

1 per cent. The results were therefore published* and the mystery of the diversity in the parallax measurements on Venus was deepened by this large difference between the radar and other measurements. However, at this same close approach the observers at M.I.T. repeated their work with improved data handling equipment and failed to find any planetary echo at the range indicated in their 1958 work. There was much unease about the validity of the previous measurements but little could be done until the next close approach in the spring of 1961.

In the meantime the small team at Jodrell concerned with the planetary radar programme explored all possible avenues for improving the performance of the system. The receiver was the easiest target for improvement and by the use of an electron beam parametric amplifier the noise figure was reduced from 4·6 to 2 db. The transmitter-feeder system was the hardest to improve. In its original form it should have delivered 100 kW in the pulse, but the feeder system was inadequate. Further, Evans wanted to increase the peak power to 200 kW in 2 ms pulses at a pulse repetition rate of 30 per second. This required a new modulator, and discussions began with English Electric at Stafford about a suitable design. In spite of their help and the replacement of various feeders on the telescope we scarcely improved this aspect of the equipment. In the end we got 60 kW in 30 ms pulses at 1 per second—a negligible improvement over the 1959 performance.

We were in trouble because we had lost J. V. Evans to the U.S.A. I was aware that he had been making a great impression on our frequent American visitors who were at that period striving hard to establish themselves in these new fields of radio astronomy. At least in those days the general courtesies were observed and contact was first made with me. Soon, a general procedure of direct offers to the individual became common practice, without reference to or interest in the effect on the organization in which he worked. However, in those early days of the 'brain drain' I first received a letter from Dave Heeschen the director of the National Radio Observatory at Green Bank in West Virginia. 6 January 1959: 'Would you be adverse to the idea of John Evans spending a year with us, on leave of absence from you? I have not yet written him about it, and have no idea whether he would be interested, but thought I should ask you first.' I replied on 12 January:

I greatly appreciate your kindness in writing to me first before approaching John Evans. I am afraid that the position in that quarter is very unfavourable in the sense that if he left us at the present time then a huge chunk of our programme would collapse. You may know that apart from his Moon work

* J. V. Evans and G. N. Taylor, *Nature*, **184**, 1358, 1959.

he is very deeply tied up with the forthcoming attempts to do Venus by radar and with our co-operative programme with the U.S. in their probe tracking. Of course Evans is a completely free man and it would not be my policy to put anything in his way if he wished to react to any offer which you might make. However, you have been kind enough to ask me for the background and as you will see we should be put in a very serious jam if he left us.

At least the threat enabled me to get Evans promoted to a more senior research fellowship, but this staved off the inevitable for little more than a year. The next approach to me from Dr. Harrington, the head of the Radio Physics Division of the Lincoln Laboratory of M.I.T. was more subtle.

29 May 1959: Recently several of our staff members have had the pleasure of visiting your laboratory at Jodrell Bank and discussing problems of mutual interest with some of your people. In order to carry on this association, we would like to invite Dr. John Evans to visit the Lincoln Laboratory during the month of June. As you know, we are planning to repeat our radar experiment on Venus this fall and this subject will be of prime interest, insofar as mutual help and co-operation is concerned.

The West Coast scientists were also on the trail. In August Professor Eshleman of Stanford asked me if Evans could attend the Conference on Radar Astronomy in October.

In addition we would like to have Dr. Evans visit for several days at Stanford to discuss scientific work of mutual interest. Both Stanford University and the Stanford Research Institute have expanding research programmes which include several studies based on radar echoes from the Moon. I would hope that all of us would benefit from intensive discussions on this important subject.

Evans returned from these two visits deeply impressed by the massive technical aid available in the U.S. for these radar programmes. Whereas he had one technical assistant at Jodrell, the Lincoln Laboratories had an army of engineers and technicians together with a transmitter vastly superior to the one at Jodrell Bank. In the spring of 1960 he resigned his fellowship with us. For me it was the beginning of a distressing series of losses of the brilliant young men who had been with me throughout the crisis of the telescope and whose devotion and skill had been a determining factor in the immediate success of the instrument. But who could expect a young man to resist a lavish red carpet reception and an offer of a salary many times greater than any sum which we could possibly offer him?

So we lost Evans at the height of our attempts to make a better system for the next attempt on Venus. Characteristically, he left copious notes and memoranda to help us on our way. On his final report, in which he calculated that if we proceeded with the system as then specified we should obtain 'a satisfactory echo in 5 hours integration, whilst the same signal to noise ratio as observed previously would be obtained in about 1 hours integration,' I find that I wrote 'Conversation with Evans before departure May 6 1960 (a) Prior (the technical assistant) fully competent on integrator, Taylor on transmitter and Thomson on r.f. also eager on parametric (b) considers programme with present equipment perfectly feasible on this basis with injection of one or two students (c) does not recommend modulator mods. unless more senior person is available.'

Evans wrote to me at the end of June. 'I am at present trying to settle down to work here at Lincoln Laboratory, and finding that the high temperatures and the security system are my greatest enemies.' We derived some consolation from his comments on the planetary radar system there which he did not think was satisfactory in its existing form. 'It would seem doubtful that the complete system will be in operation before late 1962, and even then it will be required for experiments of which I have yet to be informed.'

In September he was writing to suggest that the best system for the 1961 Venus attempt would be a bistatic experiment in which they transmitted from Millstone Hill and we received at Jodrell. I replied on 2 November 1960:

On the question of the forthcoming Venus experiment I think there is no doubt that the time would best be used by transmitting and receiving on the telescope at Jodrell Bank, and not by engaging in a bistatic experiment. As you will appreciate there is a problem of psychology involved here, and it would not be very good to change the programme when the small team left behind have been working so hard to bring about the present plans. It will indeed be extremely interesting to have this problem of the parallax finally resolved.*

It is of interest that during this correspondence we explored the possibility of installing the same type of transmitter used at Millstone Hill on the Jodrell Bank telescope. The cost was $400 000 at the

---

* During the spring and early summer of 1960 we had been a command and tracking station for the U.S. Pioneer V space probe as described in *The Story of Jodrell Bank*. The resulting orbital information on this probe had enabled another value of the solar parallax to be determined and this was in close agreement with the best of the dynamical estimates, differing considerably from the apparent radar values derived from the 1959 attempts.

manufacturers, and the estimated cost of shipment and installation on the telescope increased this estimate to $700 000. Further a 4000 sq ft building would be required and two or three skilled technicians to maintain the transmitter alone. By comparison our own transmitter and our efforts seemed amateurish, but in the event both the Lincoln scientists and ourselves were pre-empted by several weeks.

The *New York Times* of 17 March 1961 wrote: 'A space tracking station in California's Mohave Desert has established radar contact with the planet Venus, the National Aeronautics and Space Administration said today. Radar signals were bounced off the planet on March 10 by the Goldstone Tracking station.' The same day Evans sent me more detailed information which he had received by phone from Goldstone. A CW system was used with two [85 foot dishes, one transmitting and the other receiving. The transmitter power was 10 kW on 2388 MHz. The receiver used a maser with a far better noise figure than could possibly be obtained either at Millstone Hill or at Jodrell. In fact Evans said that although they had been operating at Millstone Hill since the beginning of March no success had been achieved.

The close approach of the planet was not until 11 April and we had little hope of success ourselves until near that epoch. In the event we succeeded on 8 April and continued until 25 April, by which time the Millstone Hill experiment had also succeeded. Thus, within one month, two American systems and our own were successful in the attempt to obtain planetary echoes.* The three were unanimous in the close agreement of the radar value for the solar parallax and in the repudiation of the previous discrepant results from the 1959 attempts. Although the total spread in the radar results was less than 2500 km in an average value of 149 600 000 km there remained a large discrepancy of 60 000 km between these and the best value of the conventional optical and dynamical values.†

Although we had not succeeded in increasing the power output of the transmitter significantly, our system was a great improvement on that used in 1959. The improvement in the receiver has been mentioned. In addition a more sophisticated aerial and recording system was used. In order to eliminate the fading, arising from Faraday rotation of the

* And subsequently it transpired that the Radio Corporation of America using a B.M.E.W.S. defence radar system had achieved echoes although few details were available because of the security classification. News was also received that the Russians using a CW system on 700 MHz had been successful.

† The radar value of 149 600 000 km was eventually accepted as the agreed figure for the solar parallax.

plane of polarization in the Earth's ionosphere, the primary feed at the focus of the telescope was a crossed coplanar dipole system fed through a hybrid ring, which also provided 40 db isolation between receiver and transmitter. The aerial transmitted right circular polarization and received left circular for alternate periods of about 5 minutes—the light-time to Venus and back. The receiver output was fed to an integrator which sampled at eight adjacent positions on the time base. The outputs from corresponding sampling periods on successive sweeps were counted in eight separate counting channels. Noise alone produced counts in these separate channels which were random fluctuations about the mean, a significant rise above the mean in one channel would indicate the presence of an echo at the appropriate range. A complication was introduced by the fact that the distance of the planet was changing by as much as one of these 30 ms sampling periods within a few minutes. An automatic system altered the position of the sampling channels so that an echo would always remain in the one channel. Another adjustment was necessary because the rate of change of the Earth–planet line of sight velocity caused a doppler shift in the frequency of the returned signals, which exceeded the receiver bandwidth (67 Hz) every few minutes. The receiver was tuned manually to keep it within ± 3 Hz of the predicted frequency. Whereas the 1959 attempt produced an uncertain (and as it turned out, erroneous) result after 58 hours of integration; this new equipment yielded a clear-cut and decisive answer after only a few 5 minute integration periods.*

With a decisive agreement over the value of the solar parallax, attention was now concentrated on the problem of the determination of the rotation rate of the planet. If the planet is rotating with its axis perpendicular to the line of sight with period t, then the frequency spread introduced into the reflected signal because of the differential line of sight velocities limb to limb is given by the formula on page 191. With a pulsed radar system the effect of this will be to broaden the reflected pulse. With CW working in which the intrinsic bandwidth of the transmissions can be made extremely small, an analysis of the spectrum of the reflected signal should give accurate information about the rotation rate. During the 1961 close approach attempts were made to estimate this rotation rate. Although all workers agreed that the spectrum had a narrow central feature a few Hz wide, there was strong disagreement on the interpretation in terms of the rotation rate of the planet. The M.I.T. and J.P.L. workers concluded that the planet was locked, turning one face towards the Sun. The Soviet workers, however, found a wide base in their

* J. H. Thomson, J. E. B. Ponsonby, G. N. Taylor, and R. S. Roger, *Nature*, **190**, 519, 1961.

spectrum, some 400 Hz in extent, and concluded that the rotation period was 10 days. Clearly, the enigma of the rotation rate of Venus remained to be solved.

This consideration led us to abandon the pulsed klystron transmitter in favour of a continuous wave transmitter which had been installed on the telescope by the Americans for the control of the Pioneer V space probe in 1960. This transmitter had been mounted at the base of the aerial tower and was originally hoisted into place through a large hole which had to be cut in the bowl surface. When N.A.S.A. (who had taken over responsibility from the Space Technology Laboratories) kindly agreed to let us have this transmitter on indefinite loan, we removed it from the bowl by using a helicopter. After modifications, it was re-installed for the planetary observations at the top of one of the telescope towers. The 5 kW CW power output was fed through a rotating joint at the elevation bearing and a 3 inch airspaced cable to the primary feed at the focus.

The main new technical difficulty was the extreme narrowness of the spectrum of the reflected wave which had to be anticipated. Fortunately, by this time we had surmounted the problems created by the departure of J. V. Evans in 1960. J. H. Thomson had assumed overall responsibility for the radar programmes, and in the autumn of 1960 we added to his small team a young man, J. E. B. Ponsonby, who came to us to work for a higher degree after graduating in Electrical Engineering from Imperial College. Ponsonby's later developments have already been mentioned in Chapter 8. At an early stage Thomson directed Ponsonby's attention to this problem of extreme difficulty and complexity. Essentially it was necessary to receive the signals in channels of width not more than 1 Hz. The doppler shift caused by the motion of the planet was computed to change by 35 000 Hz during the course of the observations—and at varying rates, typically of 1 Hz in 15 seconds. This doppler shift had to be removed by tuning of the receiver to at least 0·1 Hz if the 1 Hz bandwidth of the channels could be exploited.

Ponsonby solved this problem by developing a pre-programmed digitally controlled local oscillator which became known as the 'ephemeris doppler machine'. A description of this remarkable device, which had to correct the Earth–Venus line of sight velocity to better than 4 cm s^{-1} has been given in the account of the observations made during the close approach of November 1962.* Another value for the astronomical unit of 149 596 600±900 km was obtained by comparing the observed overall doppler shift of 30 kc s^{-1} during the 6 weeks of the experiment—with

* J. E. B. Ponsonby, J. H. Thomson, and K. S. Imrie, *Mon. Not. R. astr. Soc.*, **128**, 1, 1964.

the predicted values. This was in good agreement with the previous radar measurements. As regards the spectrum of the returned signal, it was possible only to place an upper limit of 1 Hz (the width of the channels) on the width between half power points. In a discussion of this result, the authors concluded that if Venus reflected similarly to the Moon (that is as a bright central disc) then this result indicated that the rotation rate was slow, probably intermediate between direct rotation with a period of 225 days (the locked case) and retrograde motion with a similar period.

The results of two further series of measurements made during this 1962 close approach were available by the time of my 1963 visit to the Soviet Union. The Soviet scientists had again been successful with their CW 700 MHz system. Their sensitivity was six times greater than in 1961 and the doppler shift was removed at the receiver with a pre-programmed local oscillator in steps of 0·2 Hz. In the returned spectrum they detected three components—a narrow spike with a width of less than 1 Hz. A component detectable to ± 8 Hz and a diffuse wide component similar to that discovered in 1961 of width 300 to 400 Hz. The origin of this was unknown; the narrower components indicated a retrograde motion of the planet with a period of about 300 days.

The other available results were from J.P.L. where the 2388 MHz CW system was used, this time with one 85 ft telescope for both transmission and reception. The spectrum analysis of the reflected signal indicated a retrograde period of about 250 days for the planet. A second method of analysis was of particular interest. In Fig. 49 if ω is the apparent rate of rotation of the planet about an axis perpendicular to the line of sight, and r the radius, then the differential doppler effect across the planetary disc will give a broadening from limb to limb of

$$\frac{4r\omega f_0}{c}$$

where f_0 is the transmitted frequency. However, a definite strip on the disc at a distance x from the centre will reflect in the frequency range

$$\delta f = f - f_0 = \frac{2x\omega f_0}{c}$$

Thus, if the analytical system is sufficiently precise to identify the power returned in a given frequency interval δf then it becomes possible to make a strip distribution of the power returned across the disc of the planet. Apart from identifying the power distribution for the reflection

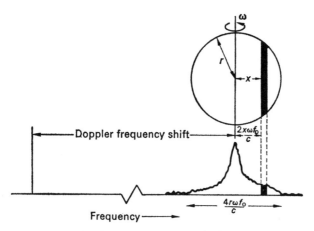

49. The effect of the spin of a planet at a rate ω on the frequency broadening of a CW signal of frequency f_0 incident on the planet. The power in the shaded part of the spectrum can be identified as coming from the shaded strip parallel to the rotation axis on the disc of the planet.

from the planet, the method should make it possible to observe the rotation of the planet if any bright strip features can be identified. In the J.P.L. observations such a bright feature was found and in 30 days of observation this feature moved 5 Hz in the spectrum. This was compatible with a retrograde rotation of 230 days ($+40$–50 days).

When I had my discussions in the Soviet Union in the summer of 1963, the position on Venus was that the problem of the astronomical unit had been settled in the radar case (but not the discrepancy with the optical and dynamical values); the preliminary indications were that the planet was in retrograde motion with a period probably lying between 250 and 300 days, and the J.P.L. identification of the bright feature had indicated the possibility of a detailed radar surface-feature investigation of the planet. The further advancement of these two latter problems (and the extension of the work to more distant planets) needed greater sensitivity than existed at that time either in the U.S.S.R., at J.P.L. or at Jodrell. We had the largest dish but the poorest transmitter. J.P.L. were not working in our frequency range. The Soviets were; they had an enormous transmitter and a reasonable size of dish, and so it seemed quite natural to suggest a bistatic experiment—the Soviets transmitting to the planet, with us receiving at Jodrell Bank.

Although it was rather easy to suggest this collaboration the problem of putting it into practice turned out to be even more formidable than I had imagined. Two months after my return I fulfilled my promise to let Kotelnikov, the head of the Institute of Radio Technics and

Electronics responsible for the Crimean installations, have details of our proposals.

19 September 1963 Lovell to Kotelnikov: Dr. J. H. Thomson who is in charge of my planetary radar group at Jodrell Bank has drawn up a short memorandum setting out the position as we see it here. You will see from this that we believe great value would accrue if it was possible to carry out this experiment, particularly on the planet Venus. We think that observations for a period of ± 50 days around closest approach would enable us to obtain more accurate values of the rotation period and the direction of the axis of rotation as well as the detection of the individual surface features.

In this memorandum Thomson worked out the improvement in performance which might be achieved in this bistatic experiment. He had the performance figures for the Soviet equipment which I had brought back with me, and these detailed calculations confirmed that the correct procedure was for us to change our 408 MHz receiver to the 700 MHz frequency used by the Soviets and for them to transmit with Jodrell receiving. This would give a 5 db improvement on the Soviet system alone, and 13 db on the Jodrell system alone. The reverse case with Jodrell transmitting and the Soviets receiving was shown to be 5 db worse than the Jodrell system alone and was not therefore discussed further. We proposed that for a period of ± 50 days around closest approach intensive observations should be made. From each day's observations we anticipated a spectrum to the limbs of the planet and a determination of the doppler shift to $\pm 0 \cdot 1$ Hz. From the variation of spectral width with time we expected to get an accurate value of the rotation period, and a determination of the direction of the rotation axis (hitherto assumed perpendicular to the plane of the observations). We also expected to detect individual surface features and make a more precise determination of the orbit and the astronomical unit. We pointed out that the planets Mercury and Mars should be readily detectable by this proposed bisstatic system.*

The memorandum stated that we would need information about the precise transmission frequency and stability of the Soviet system, and of the exact latitude and longitude of the system so that we could calculate the doppler ephemeris. I had an uneasy feeling that the Soviets might not be too willing to supply this information. Whether or not this was the difficulty which held up Kotelnikov's reply for nine months remains unknown.

---

* In June of 1962 the Soviets reported that they had obtained radar echoes from Mercury. J.P.L. reported success in May 1963. In February of 1963 both the Soviets and J.P.L. reported the successful detection of Mars by radar.

Kotelnikov to Lovell 22 July 1964* In connection with difficulties we had encountered. I could not unfortunately reply immediately to your suggestion
about joint work with Venus radar. At the present time such an experiment
might be accomplished in an alternative form, transmitting from the Crimea
and receiving at Jodrell Bank†. In our view, it is more expedient for our transmitter to work at a fixed frequency in the CW mode, and the corrector for the
displacement of frequency by the doppler effect introduced into the receiver.
If you have no objection to such a variation then please let us know when you
consider it advisable to begin work. If you have encountered difficulties of
doppler shift with the working out of your programme, we are prepared to
help you in this. It is desirable that you have a record of the received signals
provided in a suitable form on magnetic tape, which it should be possible to
send, so that we might be able to take part in the processing of the information received. The data for the Crimea station transmitter are as follows:
(1) Nominal value of frequency: 768 719 220 c/s; (2) Power density during
transmission: 250–350 megawatt/steradian; (3) polarization of transmitted
waves: circular, righthand. Astronomical co-ordinates of transmitting aerial:
latitude $\varphi = 45° \, 10'$, Longitude $\lambda = 33° \, 15'$.

After consultation with Thomson, I replied on 13 August expressing
pleasure at this favourable response and with regard to the timetable:

... our present equipment works on a frequency of 410 mc/s and it will therefore take us a little time to make the necessary modifications for reception on
768 mc/s. It will, for example, be necessary for us to obtain a new parametric
amplifier and to manufacture a new aerial system, and together with various
other modifications to the circuits we doubt if this work will be completed
until the end of this year. We therefore suggest to you that we might provisionally schedule the first experiments on Venus for the beginning of 1965,
the exact timetable to be agreed later this year when we have made progress
with our modifications.

From that moment we remained in close touch with Kotelnikov and
the endless delays which began to hinder the programme had nothing to
do with communication difficulties. Our problems were exacerbated by
the illness of Thomson.‡ On 20 November 1964, I was the one apologizing for further delays. Lovell to Kotelnikov: 'As regards the date of
commencement, I am afraid that we have not made such rapid progress
with the modifications to our equipment as we had hoped. . . . We now
propose that we work towards a target date of the early spring say some
time in March or April, when we hope it will be possible to carry out

* Translated from the Russian.

† But this was precisely the recommendation in our memorandum.

‡ Thomson relentlessly struggled against the mortal illness which led to his death
on 12 August 1969. In these intervening years his devotion to science enabled him to
sustain this onerous burden of the Venus programme and to develop the complex
concept of the lunar radar aperture synthesis.

the initial tests on the Moon.' In January 1965 I wrote that we were proceeding with our preparations and in April Kotelnikov wrote: '. . . we can start work on 1st July 1965 and carry on with it regularly in the course of the year with 2 hour contacts two or three times a week.'

This suggestion for the detailed periods of observation did not suit us at all. We had to plan our telescope programme so that the installations at the focus could work for weeks at a time continuously and then be changed to another system. On 23 April I made the counter-suggestion that 'we observe for 2 consecutive days every two weeks, and that on these occasions the observations should be carried out for the whole period of the mutual visibility of Venus on those 2 days'. On the 4th and again on the 26th of June I cabled Kotelnikov to say that we were ready to begin in July. He cabled on 2 July to say that they could not begin until the end of the month and at the same time wrote to agree our suggested schedule '. . . we will let you know an accurate date for our joint experiment later, ten days before the start.'

My difficulties in arranging for continuity of the telescope programme were now so great that I began fervently to wish that I had never suggested this collaboration. The situation was not improved by the telex message from Kotelnikov of 4 August 1965: 'With regret we can only begin the work at the beginning of October.' On 28 September Kotelnikov telexed a proposed programme testing on the Moon on 11 October and Venus from 12 to 15 October. Now it was my turn to be difficult—we could not stop another research programme so suddenly. Lovell to Kotelnikov 29 September: 'Deeply regret there are severe difficulties in operating on dates you suggest. Would you mind if we started in middle of November? If you agree please send me the proposed times.' Kotelnikov agreed and a new schedule was made beginning on 20 and 21 November (Moon test) and Venus on 21 and 22 November. On 17 November Kotelnikov telexed: 'Would work only on Moon November 21. Would inform 10 days ahead on possible days of work in December.' This started off a real communication problem. Unaware of Kotelnikov's own problems we interpreted this as implying that the only part of the schedule abandoned was the Moon test on 21 November and that the Moon on 20 November followed by Venus on 21 and 22 November, were still intended. I immediately confirmed this understanding in a letter and said we would 'Communicate by telegram on Saturday as you suggest.'

Unfortunately communication with Moscow via the telex operator on duty that weekend turned out to be a formidable task. We had, by agreement, used the code 'MOSCOW AELITA' and hitherto all such telex messages had reached Kotelnikoff without delay.* However, when on that Saturday I tried to send the simple message, 'Urgent Moscow Aelita attention Kotelnikov. No signals received from Moon 20 November. Expect signals from Moon November 21 and from Venus November 22. Please confirm', there ensued a fantasy of exchange between the international telex operator and our secretary at Jodrell which came out of the telex machine in the following form:

Op. RE YR CABLE TO MOSCOW AELITA PLEASE CFM IS
   MOSCOW THE TELEGRAPHIC OR WHAT.
Sec. MOSCOW AELITA IS THE TELEGRAPHIC ADDRESS.
Op. WELL WHATS THE DESTINATION DR IF THATS ALL THE
   TELEGRAPHIC ADDRESS.
Sec. DESTINATION PRESUMZBLY MOSCOW UWPSRDW
Op. CANT PRESUME DEAR HAVE TO KNOW DO YOU THINK
   THAT AELITA IS THE TELEGRAPHIC AND MOSCOW
   DESTINATION.
Sec. YES SORRY TO BE DENSE ABOUT THIS BUT MOSCOW
   AELITA WAS WHAT I WAS GIVEN.
Op. WELL IS IT OKAY WITH YOU TO CHANGE IT ROUND OR
   WOULD YOU CHECK TO MAKE SURE ITS CORRECT.
Sec. WELL I WAS TOLD IT WAS DEFINITELY MOSCOW AELITA
   TIC PLEASE.
Op. SPVR ERE DR RE THIS MESGE TO MOSCOW
   ACCORDING TO INTERNATIONAL REGULATIONS THE A
   LAST WORD IN THE ADDS MUST BE THE DESTINATION,
   SO IF MOSCOW IS THE DESTN CAN WE ASSUME PLEASE
   THAT AELITA IS THE TELEGRAPHIC ADDRESS?
   UNLESS OF COURSE THAT MOSCOW ALITZ E E E
   UNLESS OF COURSE THAT MOSCOW AELITA IS A NAME
   OF FIRM IN THAT CASE WE CAN PUT MOSCOW ON THE
   END READING MOSCOW AELITA MOCOW?
   MOM PSE
Sec. MOSCOW IS DESTINATION. ACTUAL ADDRESS IS
   INSTITUTE OF RADIOTECHNIQUES AND ELECTRONICS
   PROSPECT MARSK MOSCOW I THINK IT WOULD BE OK
   TO WRITE AELITA MOSOCW.

* The explanation of the origin of this telegraphic address was given in *Literaturnaya Gazeta*, 1 February 1966, in an article by Irina Radunskaya describing our collaboration: 'The telegraph address of the U.S.S.R. Academy of Sciences, Institute of Radio and Electronics was thought up by its workers quite by chance, and its "invention" has a humorous side to it. The Institute sent to telegraph authorities a list of five words to choose as its address. The five words proved already chosen by other organizations. Then the scientists suggested the names of five planets; they too were taken up. In the third series of five words, only one proved "free"—Aelita. [The translator adds the explanation that this is the name of a character in a sci-fic novel of the same title by Alexei Tolstoy.] That has been a happy choice, for the scientists are working on fantastic things.'

Op. FINE TKSVM IF U ARE SATISFIED THAT AELITA IS THE
REGISTERED ADDRESS FOR THE ESTABLISHMENT UVE
JUST MENTIONED WE WILL SEND IT TO AELITA
MOSCOW THEN OK WIV U OR DO U PREFER US TO PUT
THE FULL ADDS AS QUOTED?

Sec. WELL WE SENT A CABLE TO MS EE MOSCOW AELITA
ON NOV 19 AND IT GOT YHERE OK

Op. FINE ILL LOOK THAT ONE UP DR AND SEE HOW
EXACTLY IT Q WAS SENT U SENT IT THU US EH?
YES LOVELY WE ARE ONLY QUERYING THIS OF MOURSE
CAUSE WE WANT THE CABLE TO GET THERE FOR YE
WILL DO BEST TKSVM CHEERS.

I do not know whether that message ever reached Kotelnikov. We never got the signals either from the Moon or Venus. Thomson and Ponsonby were maintaining with increasing vigour that this was not the kind of research they wanted to do at Jodrell Bank—and after that Saturday telex experience I began to have doubts myself. However, after a few days interval I thought I would try once more to arrange a schedule. On 25 November I wrote to Kotelnikov to report that our apparatus worked well but we did not receive signals and asked him for some December dates, 'but we must ask you to avoid December 24, 25, 26. We cannot operate on those days because it is the Christmas holiday in England and they are the only days of the year on which we find it impossible to operate the telescope.' He replied on 29 November apologizing that 'circumstances connected with observation of the cosmic stations Zond 3, Venus 2 and Venus 3 have prevented us beginning the joint experiment at the very last moment.' He suggested a December timetable for the next attempt, which was twice altered, but on 21 December we did at least have bistatic contact via the Moon, but failed on Venus—as it turned out, because of a misunderstanding about the procedures for introducing the doppler frequency shift corrections. At last on 9 January 1966 we were able to cable Kotelnikov that strong signals had been received via Venus.

From that date a regular sequence of measurements, on a schedule agreed by telex, proceeded without further problems, apart from the increasing weariness of Thomson and Ponsonby who were bearing the main load of this work at irregular hours of the day and night.

15 February 1966 Lovell to Kotelnikov: In about one or two weeks time we hope to begin the despatch to you of the magnetic recordings in the form requested in your letter of 2 July 1965. . . . At this stage it seems to us to be desirable to consider again the objectives of the experiment. In the memorandum prepared by Dr. Thomson in August 1963 paragraph 5 listed two objectives.

(a) Observation for a period of ± 50 days around close approach leading to a determination of the rotation period of Venus, direction of axis, the astronomical unit and orbital data.

(b) Away from close approach; orbit and astronomical unit determination.

Dr. Thomson and Ponsonby who are running this research programme feel that the observations needed to meet the first objective will be completed in the near future. . . . We are now inclined to the opinion that, perhaps, after a few more sessions it might be desirable to suspend the observations so that Dr. Thomson and Ponsonby can have sufficient time to do some analysis and also collect the data for study by you and your staff.

There is no doubt that objective (a) will have been attained, and we should then take the opportunity of this intermission to define any future scientific objectives of a renewal of the programme. It is obvious that many new results have been obtained by direct radar measurements in the U.S.S.R. and U.S.A. since the 1963 memorandum was written.

Indeed we began to feel increasingly that this programme, originally proposed in 1963 for the 1964 approach and now only taking place in 1966, was already overtaken by other events. During the June 1964 close approach we had ourselves repeated the previous measurements with a further improvement in sensitivity which yielded more precise spectral data.* With still further refinement the American workers were also beginning to obtain more exact details of the rotation rate and some topographical detail of the surface. The immense Arecibo dish in Puerto Rico, located so that the planets passed through its beam, showed so much increase in sensitivity over all other systems that its superiority soon became a great discouragement to other workers.

However, the sole response to my letter was a continuation of the telex messages setting up the schedule for the obervations, and the data continued to accumulate. Finally, on 18 March Ponsonby and Thomson could stand the strain no longer. They deposited a memorandum on my desk listing ten good reasons for ceasing the measurements, not the least conclusive being that either one or both would be away for Easter during the next month. Three days later another of the regular scheduling telex messages arrived for the 28 March operation which elicited this response from me:

Lovell to Kotelnikov 21 March 1966: Thank you for your telegram suggesting next schedule of operations but have today despatched to you a letter to say that we consider sufficient results have now been obtained to complete first stage of experiment. Letter suggests a discussion of points raised in my letter of February 15 before making decisions about next operation particularly an investigation of frequency discrepancy. We hope to send you information about this in a few days.

* J. E. B. Ponsonby, J. H. Thomson, and K. S. Imrie, *Nature*, **204**, 63, 1964.

The reference to the frequency discrepancy was to a problem which was causing us much dismay. In the first response which Kotelnikov made to our memorandum in July 1964 he gave the frequency of the transmitter as 768 719 220 Hz. However, when we finally obtained the signals reflected from Venus, we found that the frequency transmitted by the Soviets must be 30 Hz lower than this value, and, moreover, it varied from session to session in the range 25 to 35 Hz low, and on one occasion 39 Hz low. It was, of course, essential to clarify this issue before any worthwhile analysis could be made of the reflected spectrum, but the Soviets seemed particularly reluctant to comment on this matter. The obvious conclusion was that the frequency standard used at the transmitter was remarkably poor by contemporary standards. Thomson and Ponsonby pressed me for any recollections which I had about this during my visit. The recollection was a vivid one, but not very helpful, namely that in the underground rooms containing the transmitter there was one room said to be the frequency control standard which was actually guarded. As with so many of the Russian puzzles, the answer was probably simple, but we never found out the answer to this one.

My letter to Kotelnikov in further explanation of the telegram re-iterated the points made in my earlier letter of 15 February but I felt bound to place more emphasis on this frequency problem. 'We are also particularly worried about the discrepancy in the frequency and we can-not find any reasons for this phenomenon. It would in any case not be realistic to continue with the second part of the programme until the reasons for this anomaly can be investigated and understood.'

Our telex messages seemed to reach Kotelnikov and elicit quick responses whereas our letters did not. At least there was a cabled response to the 21 March message agreeing to 'an interval' in the observations. Though we were beginning to have reservations about the scientific value of continuing this experiment in the light of the new results now available from the U.S.A., nevertheless the liaison had been a happy political event. At this time the Scientific Attaché in Moscow sent me a glowing account of our collaboration by Irina Radunskaya published under the title 'A mirror for Venus'.*

Eventually, in August of 1966, with diplomatic help, the despatch of the magnetic tapes to Kotelnikov began. Seven tapes from which Ponsonby had extracted spectra were delivered to Kotelnikov via our Embassy in mid August. On 2 August I had written to him about these arrangements. My letter continued:

As you may know there is a joint session of Commission II and Commission V at the U.R.S.I. Assembly in Munich in September to deal with planetary

* *Literaturnaya Gazeta,* 1 Feb. 1966.

radar, and it seemed to me that it would be an excellent idea if some mention could be made of this joint experiment. I have therefore suggested to the Chairmen of the Commissions that an opportunity should be made for a few minutes presentation of this experiment and an indication of the results. Mr. Ponsonby hopes to show a spectrum of this type (enclosed) and I hope that if you are at U.R.S.I. you would also say something about your aspects of the experiment, or if you are unable to be there yourself that one of your representatives could do so.

Kotelnikov replied by cable 'Thank you for magnetic tapes which we received. Unable to take part in U.R.S.I. General Assembly. Agree you present a preliminary communication of our joint experiment.' Ponsonby did present an appropriate communication in Munich, but the lack of explanation from the Soviets of the frequency problems and the absence of the Soviet scientists detracted from the occasion.

The remaining magnetic tapes were transmitted at regular intervals during the following months and by mid March of 1967 the consignment—amounting to 80 per cent of all the tapes recorded—had been completed. On 27 June Kotelnikov sent me an example of their own analysis of the spectra and wrote that he intended to present an account of the results at the I.A.U. Assembly in Prague in August. On 2 August I informed Kotelnikov that Dr. Thomson would be at this meeting: 'We are, of course, quite agreeable to your presenting the results of this work at the I.A.U. meeting in Prague and I hope you will discuss with Dr. Thomson the possibility of making a joint publication later on of the results.'

Unfortunately, conclusions which would have been exciting if the work could have been carried out when first suggested in 1963, had by 1967 been published by the Americans and enthusiasm for further publication evaporated. A short description of the work with an example of the power spectrum obtained as a result of $2\frac{1}{2}$ hours integration on 27 January 1966 was published in the 1968 edition of the Soviet Year Book *Science and Humanity*.* Ponsonby's detailed analysis of the work formed part of his Ph.D. thesis.† He exhibited great ingenuity and skill in the computational analysis to produce the type of spectra shown in Fig. 50. This is a copy of the line-printer output of the Atlas computer for 9 February 1966. But his conclusions on the rotation rate and preliminary location of surface features of the planet had already been pre-empted by the American publications.

---

* As part of my article on 'The Radio Astronomical Observatory at Jodrell Bank, England.'

† J. E. B. Ponsonby, 'Planetary Radar', Thesis for the degree of Ph.D. University of Manchester, 1969.

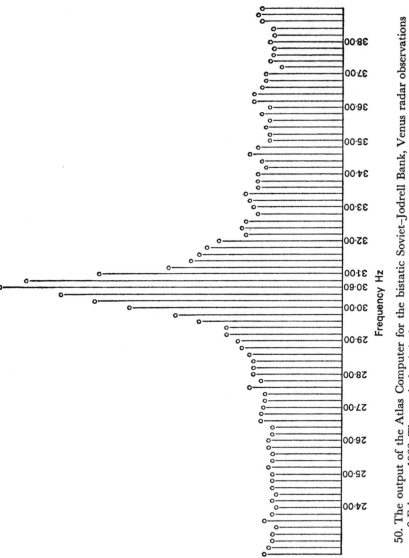

Frequency Hz

50. The output of the Atlas Computer for the bistatic Soviet–Jodrell Bank, Venus radar observations on 9 February 1966. The vertical axis is the strength of the signal reflected from the planet (in arbitrary units). The horizontal axis is the frequency in Hz, each vertical line separated by 0·2 Hz and corresponding to a strip distribution as shown in Fig. 49.

15

Tripartite collaboration with the U.S.A. and U.S.S.R. via the Moon and balloon

In Chapter 14 I referred to our early work on radar reflections from the Moon. When the Mk I telescope came into use in the autumn of 1957 it was an easy matter to establish a transoceanic voice link using the Moon as a reflector. These experiments were of practical interest. In those days communication satellites were little more than a dream and there was a real problem about radio communication over long distances because of the black-outs and fading which arose when solar disturbances affected the Earth's ionosphere. This happened because the ionosphere itself was used to reflect the radio waves for transoceanic radio communications. An obvious solution was to use the Moon as a reflector with radio frequencies high enough to penetrate the ionosphere so that they were not affected by such disturbances. The difficulty was that, because of the fading, it was not considered that a sufficient bandwidth could be realized to make intelligible voice transmissions feasible. J. V. Evans, as part of his postgraduate degree research, made a detailed study of the short-period libration fading. He concluded that on the frequency which he was using, 120 MHz, the Moon was not behaving as a bright disc, nor was it a reflector obeying the well-known Lambert scattering law (with a cosine distribution). On the contrary, Evans's analysis showed that the characteristics of the rapid fading could be explained only on the assumption that, on these frequencies, the Moon was limb-dark and that the effective scattering area was a region at the centre of the visible disc, having a radius of about one-third that of the lunar radius.

This surprising result, which Evans soon verified by more direct measurements of the broadening of the reflected pulse, led to the prediction that the Moon might be used in a communication circuit for intelligible speech. In the autumn of 1958 I told Evans I wanted to demonstrate this in my Reith Lectures, which I was about to deliver.

We suspended ourselves in the laboratory underneath the bowl of the Mk I as it was tracking the moon. Our 'hellos' spoken into a microphone there were clearly received $2\frac{1}{2}$ seconds later after reflection from the Moon. This demonstration formed a part of my third Reith Lecture.*

The experiment was made with a small transmitter of low power on the telescope, but it served to stimulate the interest of one of the directors of Pye Telecommunications Ltd. who was listening to the lectures. We were soon extending this work in close collaboration with them, using a more powerful and appropriate transmitter of their manufacture. Within a short time clear voice circuits via the Moon had been established between Jodrell Bank and an American telescope, and it was even found possible to transmit recognizable music. Finally we achieved a brief contact with our Australian colleagues via the Moon circuit—although of course in this case the requirement of mutual visibility of the Moon at the transmitting and receiving ends placed a severe restriction on the possible periods of communication. In spite of this restriction to mutual Moon visibility, and of the $2\frac{1}{2}$ second delay occasioned by the distance of half a million miles which the signals had to travel, it seemed likely that the Moon communication system would have a considerable commercial future. However, before the plans of Pye Telecommunications Ltd. to use small 25 ft dishes and somewhat more powerful transmitters, developed practically, the possibility of using Earth satellites emerged. The successful use of the initial low orbit satellites, and later of the stationary orbit satellites removed the commercial incentive for the Moon communication link. The system was, however, subsequently widely used for military purposes.

The Earth satellites whose use soon became standard practice for communication contained receivers and transmitters. Before these techniques developed, the concept of using a large balloon, its surface coated with a metallic reflecting layer, was tried. The idea was quite simple—by launching a small number of such balloons into suitable orbits around the Earth continuous worldwide coverage could be achieved by reflection from these artificial moons.

The *Project Echo* of 1960 was N.A.S.A.'s first major experiment in a programme to test the feasibility of using a passive satellite reflecting sphere for long-range communications. During the early spring of 1960 it was intended to launch into a nearly circular 900 nautical mile orbit a 182 pound payload consisting of a container 28 inches in diameter. The deflated 100 ft diameter reflecting sphere was folded inside this container. The idea was that after injection, the container would separate from the third stage and after a suitable interval the two halves of

* *The Individual and the Universe* (London 1958).

the container would be separated by a small charge. The balloon, made of $\frac{1}{2}$ mil thick aluminium coated mylar, would then be exposed to the vacuum of interplanetary space and would inflate under the influence of the sublimation of 10 pounds of benzoic acid placed within the sphere.

The primary experiment was to be an attempt to establish communication via the balloon from the Bell Telephone Laboratories in New Jersey to the Goldstone Tracking Station in California, but N.A.S.A. were anxious to establish communication across the Atlantic with Jodrell Bank. On 3 March 1960, Bill Young, the head of the U.S. team then at Jodrell Bank for the Pioneer launchings, handed me a long telex message. 'N.A.S.A. . . . have solicited their [Jodrell Bank's] assistance. However, they have received very little information from the station and have assumed from this that they are not very much interested.' On 7 March I replied to Dr. L. Jaffe, the chief of Communication Satellite Programs at N.A.S.A.: '. . . if N.A.S.A. desires our participation in the communication experiments we will do so provided we can have assistance from S.T.L. personnel stationed here. We understand S.T.L. would be willing to co-operate provided there is no interference with the Able-Thor programme.'*, to which the reply came three days later that N.A.S.A. would 'greatly appreciate' our participation.

In this way, when we were much preoccupied with the Pioneer space launchings, we assumed a special relationship with the Echo balloon project, which, in a few years was to lead us unexpectedly into a fascinating attempt at Soviet–Jodrell Bank–U.S. collaboration. However, in the spring of 1960 those further complications were not envisaged, and the discussion of our collaboration involved only receiving and transmitting signals via the balloon between Jodrell Bank and New Jersey. By early April we had received the projected orbital data and had agreed to respond to N.A.S.A.'s request to provide radar tracking data for the first two or three days, in addition to the participation in the communication tests. The complication of dovetailing these arrangements with our own researches was exacerbated by the vagueness about the date of launching. On 30 March a message from the Director of Space Flight Programs informed me that the launching would be on 28 April but—'Please hold this date private until launch'. On 21 April a circular letter to COSPAR from the U.S. Space Science Board, said that the launching would take place during the first or second week of May. On 26 April, Space Flight Operations informed me that the launch would

---

* S.T.L.—that is the Space Technology Laboratories of Los Angeles, originally contractors to the U.S.A.F. for the Pioneer (Able-Thor) Moon and space programmes. See *The Story of Jodrell Bank*, pp. 237, 240. At that time N.A.S.A. had not yet assumed control of all the U.S. space programmes.

be on 5 May 'with daily postponement dates through May 15', and

... with regard to relative priorities between tracking Echo 1 and participating in the communications experiment, we request that during the first day
you give full priority on all passes to radar tracking. During the next four days
we suggest that you try receiving Bell Telephone Laboratories' signals on the
most favourable communications pass of each day and track on the remainder.
After the fifth day, the orbit should be sufficiently well established that
further tracking should be unnecessary.

Since our radar tracking had to be on 100 MHz and the receiving equipment was on 960 MHz aerial changes were necessary, but at that time
our arrangements were still simple and in an internal note about the
schedule I remarked that 'we are doing the radar tracking on the passes
preceding and following number 7, but since there is a period of 2 hours
between passes there will be adequate time for the interchange of the
aerials'.

In fact, our preparations were to be of no avail. After several post-
ponements, the launching was attempted on 13 May. The next day I
received the following message from N.A.S.A. 'It now appears to be no
possibility that Echo achieved orbit. Further search unjustified.' Although, three months later, on 12 August 1960 a second attempt was
successful, the Jodrell Bank telescope was completely out of action. In
early July, immediately after the completion of the deep space Pioneer V
operation, urgent action was taken to replace the cracked ballbearings
in the central pivot.* On 22 August, ten days after the successful launch
of Echo I, we were in action again and for three nights we were able to
carry out the tests on suitable orbits of the satellite. Our participation in
the Echo I experiment transpired to be without great significance. As a
result of the comprehensive tests between New Jersey and Goldstone the
Jet Propulsion Laboratories concluded that

... unless the communications circuit gain can be increased by at least 30 dB
over that used in Echo I, transoceanic television will not be possible, even as
an experiment. Also, to increase the mutual visibility time per satellite, which
would be essential for a practical system, the balloon should be at a minimum
altitude of 2000 miles, which means at least another 12 dB, or a total increase
of 42 dB. If 1000 ft balloons could be launched ... then television might be
relayed on a reasonably practical basis ... other satellite techniques may
prove more suitable for the transmission of wideband video signals.†

* In *The Story of Jodrell Bank* (p. 243), I wrote that the Pioneer V episode had been
'fraught with a terrible anxiety: shortly after launching the telescope cracked a bearing in the central pivot ... we continued day by day to track Pioneer with an explosive noise like machine gun fire occasionally emanating from the central pivot ...
it was a relief when, without any danger to a vital mission, we could remove the
offending pivot and deal with the trouble.'

† Jet Propulsion Laboratory, Calif., Technical report No. 32–59, 1 Dec. 1960.

With the development of the active repeater communication satellites it seemed unlikely that the telescope would be further involved in passive communication experiments via the Moon or balloons. However, the attempts by the Americans to establish an effective liaison with the Soviets in space research involved us unexpectedly four years later in a memorable politico-scientific exchange.

On 4 March 1963 I wrote to the director of Pye Telecommunications Ltd. who had so generously participated in the earlier Moon communication experiments

. . . on the question of the Pye equipment which we had for the Moon work. A long time ago when this was installed on the telescope and when we were then using it on Echo I, we independently promised both the Russians and the Americans that we would be glad to co-operate in any future joint work between Russia and the West. Matters have now ground on and in the light of recent discussions at Geneva and elsewhere it appears that a joint experiment is now being planned in which we are to be the intermediate station. This is all being done officially with the Post Office and D.S.I.R., and I have told them that we will do our best to help. The only facts of the situation known to me are that it is proposed to carry out the experiment on Echo II which the Americans hope to launch towards the end of this year and that the Russians will be prepared to do the experiment on the frequency which we must have given them as that in use on the original experiments (namely 160 MHz or thereabouts). . . . I hope, therefore, that you will be kind enough to help again with the transmitter and receiving equipment on the telescope.

Only many years later when I read the official U.S. account* did I understand how a casual offer of help had materialized into an international agreement involving us without further consultation—and on a frequency which by that time we would have desired to change because all the original equipment had long since disappeared. During the first year of N.A.S.A.'s existence it was natural that the officials should make efforts to collaborate in space with the Soviets and during November 1959 the Deputy Administrator of N.A.S.A., Dr. Hugh L. Dryden had talks with Academicians Sedov and Blagonravov. The Soviets were shown the work in progress on the Mercury capsule but apparently showed 'little interest', and no comparable invitation was extended for an American to see the Soviet work.† Nearly two years later (August 1961) N.A.S.A. tried again. Shortly after Echo I had been launched the International Astronautical Federation assembled in Stockholm. The official record* reports that

* *Soviet Space Programs 1962–65*, Ch. V (by Joseph G. Whelan)—Committee on Aeronautical and Space Sciences, U.S. Senate, Washington, 30 Dec. 1966.

† In retrospect, this is not surprising in view of the disparity which then existed between the U.S. and Soviet preparations for manned space flight.

As delegates to the Conference were observing the overflight of Echo I, Mr. Frutkin (of N.A.S.A.) who was standing next to Blagonravov suggested that the Soviet Union and the United States communicate with each other by means of Echo I as a gesture of friendliness and a step towards co-operation. The Soviet academician agreed. But after further inquiry it was evident that the satellite was not high enough to provide mutual visibility for communication purposes between the Soviet Union and the United States. *It was then agreed that the Jodrell Bank radiotelescope in England would be used as a terminal.* The matter was left with the Americans to make the arrangements.*

In fact, nothing happened immediately because Blagonravov shortly afterwards informed N.A.S.A. that 'difficulties had arisen'. Indeed the attempts at Presidential level to establish co-operation received another setback when President Kennedy brought up the subject of space collaboration with Mr. Khrushchev at the Vienna meeting in June 1961. Khrushchev responded that co-operation was impossible because he preferred to maintain the veil of secrecy around Soviet rocketry.† Immediately after this meeting the political problems over Berlin monopolized U.S.–Soviet relations, but as the crisis dissolved so the Soviet attitude to space collaboration began to change. On 27 November 1961 they ended their boycott of the United Nations Committee on the Peaceful Uses of Outer Space. In December the Soviets supported Resolution 1721 (XVI) which established the terms of reference for space co-operation.

Under the umbrella of Resolution 1721 (XVI) the special circumstance which involved the Jodrell Bank telescope emerged. On 21 February 1962 Mr. Khrushchev congratulated President Kennedy on the success of Col. John Glenn's three-orbit spaceflight the previous day. He referred to the advantages of pooling resources in space. President Kennedy responded appropriately and declared that he had instructed the appropriate officers in the U.S. Government 'to prepare new and concrete proposals for immediate projects of common action'. Urgent action followed and two weeks later (7 March 1962) President Kennedy sent to Mr. Khrushchev American proposals for co-operation in five activities the fourth of which was 'An invitation to the Soviet Union to participate in exploring the feasibility of intercontinental communications satellite systems'. In his response of 20 March Khrushchev said that the Soviet representatives on the U.N. Committee would be instructed to discuss with U.S. representatives 'concrete questions of co-

---

* These are my italics, emphasizing the extraordinarily informal way in which the Soviets and Americans initiated the exchanges which eventually tied us in to support a formal international agreement.

† Theodor C. Sorensen, *Kennedy* (New York 1965), p. 529.

operation in research'. He listed six problems the first of which was 'communication satellites'.

As an immediate consequence of these exchanges the first meeting between Dr. Dryden and Academician Blagonravov took place at the United Nations in New York from 27 to 30 March 1962. Frutkin has reported* that Blagonravov opened the meeting by stating that he 'first wished to discharge his 2-year-old promise to the author by arranging for mutual tests by means of the Echo satellite'. But Echo I had deteriorated and the future Echo II was suggested as an alternative prospect. The Dryden–Blagonravov discussions were resumed in Geneva from 29 May to 7 June. On 8 June the first Dryden–Blagonravov agreement was reached. It provided for collaboration in meteorology, geomagnetic surveys and satellite telecommunications, the last by means of the proposed Echo II balloon satellite. On 29 August the State Department notified the Soviet Foreign Ministry that the recommendations were acceptable, the Russians responded similarly on 12 October. Subsequent exchanges between the President of the Soviet Academy, Academician M. V. Keldysh and the Chief Administrator of N.A.S.A., Mr. Webb, led, by 30 October, to the formal instruction to Dryden and Blagonravov to convene the necessary working groups. It is remarkable that these formal exchanges were completed without serious delay during the Cuban missile crisis.

During March 1963 in Rome, and May in Geneva Dr. Dryden and Academician Blagonravov, with their advisers, drew up the 'Memorandum of Understanding between N.A.S.A. and the Soviet Academy of Sciences'. Part III of this memorandum was concerned with the arrangement for passive communication satellite experiments. Both countries agreed to participate using the Echo II satellite which N.A.S.A. expected to launch prior to mid-1964. This part of the memorandum contained information on the characteristics of the satellite, frequencies, arrangements for using Gorki State University facilities, and the facilities of the University of Manchester at Jodrell Bank, the supply of orbital information, planned types of transceivers and exchange of observational data.

I first learnt of this arrangement made by the Americans and Soviets for the use of our telescope through a newspaper reporter, and this accounts for my vague reference in the letter to the director of Pye Telecommunications quoted earlier that 'it appears that a joint experiment is now being planned in which we are to be the intermediate station'. However, it appears that protocol was satisfied in that some form of official contact must have been made during the Rome discussions with

* A. W. Frutkin, *International Co-operation in Space* (New York 1965), p. 94–5.

the D.S.I.R. and the Post Office. At least on 11 March I wrote to Francis at D.S.I.R. 'A short time ago you phoned me about our possible assistance in the proposed international collaborative experiment on Echo II, and subsequently Captain Booth (of the Post Office) also spoke to me about this. In both of these conversations I confirmed that we would be willing to assist in this programme'.

The Managing Director of Pye Telecommunications Ltd. replied to my letter of 4 March instantly to convey the unfortunate news that 'the equipment which we previously lent you has been disposed of. We would be happy to try to make other equipment available but before committing ourselves on this we would like to know what precisely you want, and when.' The answer to the last query was, apparently, that Echo II would be launched before the end of the year, or at least early in 1964, and the problem of recreating the arrangements made for Echo I on the telescope suddenly assumed rather alarming proportions. Our working contact in N.A.S.A. was to be the same individual as for Echo I, Dr. Leonard Jaffe, the Director of Communications System Office of Applications, and on 1 April 1963 I sought his advice and help.

If it is decided to proceed . . . we would, of course, be glad to assist, but it is necessary that early thought should be given to the problem of the transmitter to be used on the telescope at Jodrell Bank. The one which we used previously, on which I believe the Russians have based their plans on a frequency of about 160 MHz, was borrowed from Pye Telecommunications Ltd. . . . although in principle they would be glad to help us they have actually disposed of this particular transmitter and have no other available. I would therefore be glad to hear from you about this specifically as to whether a final decision has been made on the operating frequency and also whether N.A.S.A. could supply us with the appropriate transmitting equipment to use on the telescope.

Jaffe replied on 12 April to the effect that the official discussions 'may result in a firm agreement in approximately 60 days'. He confirmed the intention to use the telescope on 162 MHz and said that the understanding was that 'N.A.S.A. will request the use of U.K. facilities necessary to these experiments through our co-operative partner on communication facilities in the U.K., namely, the General Post Office'. Although we had maintained friendly relations with the Post Office on frequency allocations throughout our history, this was clearly to be a relationship of a different order—and indeed the formal establishment of this liaison was to be more essential to us than I imagined at that moment. On the problem of the apparatus with which we were to do this work, Jaffe could only offer to visit us to discuss it and meanwhile to look into the availability of such transmitters in the U.S.

Simultaneously the following telegram was handed to me:

Blagonravov to Lovell: Request your consent to receive two Soviet experts participating Echo two experiment for mutual acquaintance with technical characteristics of radio receiving and transmitting equipment. Our specialists planning to leave for England late April if confirmation received.

Since we had no apparatus, and especially in view of Jaffe's letter, I deemed it expedient to stall and consulted J. H. H. Merriman at the Post Office, the person on Capt. C. F. Booth's staff who was responsible for handling the detailed arrangements.

19 April 1963 Lovell to Blagonravov: . . . we will be happy to discuss with your two experts preliminary technical arrangements for passive communication tests using Echo II. Regret it will be inconvenient to hold such discussions until late May. I suggest a three day visit starting May 21 to include preliminary discussions in London and a visit to Jodrell Bank. If you agree will you please inform Merriman, Assistant Engineer in Chief, G.P.O. London, who is co-ordinating the arrangements for the tests.

29 April Blagonravov to Lovell: Your suggested [date] is convenient for our experts. Drs. Herman Ghetmantshev and Igor Zhulin will be in London by 20 of May.

A last minute appeal to me by Ghetmantshev for help with their visas was dealt with and the meetings between the Russians, the Americans, G.P.O., and J. G. Davies and myself occurred without incident in London and at Jodrell Bank. At last, nearly four years after the initial approach by Dryden to Blagonravov about space collaboration, the practical details of the first U.S.–U.S.S.R. ventures were being settled—and we were in the middle of the sandwich.

At these meetings for the first time we heard of the proposed arrangements in the Soviet Union for the reception of the signals. A steerable radio telescope of 50 ft diameter at the Zimenki Radio Observatory in Gorki was to be used. The distance between Gorki and Jodrell was 1850 miles. With the balloon at an altitude of 700 miles the total path length via the balloon at the most favourable point would be 2400 miles. The small size of the receiving telescope which the Russians proposed to use somewhat damped our optimistic view that the experiment would be a simple one. In fact the detailed calculations which were made indicated that there would be adequate signal for telegraphy, but that voice communication would be hardly intelligible unless it was time-expanded.

In the meantime we had the crucial problem of procuring a transmitter on this frequency, and further negotiation with the Pye Telecommunications Company revealed that they had a suitable instrument at their branch in Ireland which we could purchase but not borrow. The commercial enthusiasm evident a few years before had unhappily, but not unnaturally, vanished. I appealed to Francis at D.S.I.R. since he had been involved in the initial contacts with N.A.S.A. earlier in the year. D.S.I.R. would be willing, in principle, to make us a grant to purchase the transmitter but the difficulty was that their grants were for the initiation of research of 'timeliness and promise'. It would be difficult to argue that this international communications experiment via the balloon came into that category, and for some time it seemed that this attempt at the implementation of U.N. Resolution 1721 (XVI) might flounder for the want of a few thousand pounds. Fortunately we devised a useful way out of this dilemma by arguing that the transmitter would afterwards be useful for our lunar investigations. So, on 17 May 1963, I made a formal application to the D.S.I.R. for money to purchase a

1 kW transmitter . . . on 162·4 MHz. . . . this equipment will be used in the first place for the international communication experiment with Echo II, nevertheless it would be of considerable scientific use when this experiment has been completed. The provision of a 162 MHz transmitter would facilitate a useful extension to the two-dimensional lunar aperture synthesis programme, now in its early planning stages after a successful preliminary one-dimensional experiment. . . . In this work the principles of aperture synthesis . . . are being applied to radar astronomy to achieve the resolution of less than one minute of arc which is needed for detailed radar mapping of the Moon.*

Rapid and favourable consideration was given to this request and well before the end of the year we had all the necessary equipment ready for these experiments.

Our problems lay in other directions. Naturally, anyone wishing to transmit has to obtain authority from the Post Office and the restrictions are quite severe otherwise there would be chaotic interference in the radio wavebands. In the normal situation we are the people who suffer because of the great sensitivity of the telescope. On the occasions when it was used as a transmitter we had followed the appropriate procedures to obtain the necessary permission from the Post Office. This had certainly been done with the earlier transmissions involved in our radar work on the Moon, and it had not occurred to me that difficulties might be made now, particularly in view of the fact that the Post Office was the official liaison body with N.A.S.A. I was, therefore, surprised to

---

* The transmitter was, indeed, subsequently used in this research with great success. The principle and theory of the research was described in 1968 by J. H. Thomson and J. E. B. Ponsonby, *Proc. R. Soc. London*, Ser. A, **303**, 477, 1968. After Thomson's death, Ponsonby carried on with the researches using the 50 ft polar axis telescope at Jodrell Bank. Some preliminary results of the high definition maps of the moon made on the 163 MHz frequency compared with subsequent measurements on 408 MHz were presented at a lunar conference in Greece in 1971. The full results are not yet (1972) published.

receive a letter from the Radio Service Department of the G.P.O. in October 1963: 'I am looking into the possibility that the proposed tests, in which Jodrell Bank will transmit on 162·4 MHz may cause interference to some radio installations, both private and Government.' The letter then asked for technical details and ended 'If it is possible to say at this stage what will be the duration of each transmission, how many transmissions are likely to be made each day and over what period the experiment will continue, that information also would be useful.' In my reply of 15 October I gave the technical information and added 'I am afraid it is not possible for me to answer your queries with any degree of precision because we do not yet know the orbit of Echo II neither has a time schedule been discussed.' The correspondence continued in a helpful and friendly manner but on 6 November under pressure because the G.P.O. felt that 'some Government radio services could be adversely affected.' I wrote that 'At the present moment we have no further information of any kind and it is now obvious that the experiment which was planned for the end of this year cannot possibly take place until 1964.'

Indeed it was clear that the progress anticipated in our May meetings was not being made and that the reasons were not technical ones. On Christmas Eve, F. J. D. Taylor of the Post Office Engineering Department wrote to me.

While visiting Washington recently I had the opportunity to discuss with Dr. L. Jaffe of N.A.S.A. the projected tests using the Echo II balloon satellite. I gathered that the satellite will be launched late in January 1964. N.A.S.A. has not so far received any reply from the U.S.S.R. Academy of Sciences to the questions posed in the record of the U.S.A./U.K./U.S.S.R. meetings held in London and Jodrell Bank last May. Recently N.A.S.A. received from Academician Blagonravov an apology for delay and promising early reply.

Obviously we could not be expected to install the equipment and stand by on this basis.

30 December 1963 Lovell to Taylor: . . . we have all the equipment here, but of course, we shall require some warning if we are to engage in these experiments. The only thing I can do at the moment is to programme the telescope on the basis of our internal research programmes, and quite apart from my desire not to make any interruptions in these programmes, it would certainly take us several days to carry out the installation of the new aerial and tests which will be necessary before we can work with Gorki.

At last on 25 January 1964 Echo II was launched—a reflecting balloon similar to Echo I but 135 ft in diameter, in a nearly circular orbit at an altitude of about 750 miles. The real difficulties then began. Initially all seemed well on the political front. Two days after the launching Blagonravov sent me a long letter which (translated) opened: 'In connection with the launching of Echo II we now have the opportunity of coming to the fulfilment of the preliminary arrangement reached in 1963 between you and Dr. G. G. Ghetmantshev on the receiving of signals via Echo II which will be sent by your observatory Jodrell Bank and received by the Radio Astronomical Observatory of Gorki University.' He then proposed a programme for the tests. 'The experiment is divided into two stages: the first stage begins not earlier than 15 days after the launching of Echo II at concrete times according to the agreement with the English side and designed approximately for a monthly period: the second stage will be carried out according to an additionally agreed programme after establishing the workable possibilities of the separate types of transmissions also over a month.' Blagonravov then listed the various types of transmissions which should be used. He asked for the 'organization of a direct telegraph communication between the observatories at Gorki [Zimenki] and Jodrell Bank', specified the detailed timetable, dealt with the agreements about communication of results and ended: 'As regards the question of the establishment of a direct telegraph link between Gorki and Jodrell Bank, ... recommend you to propose to the Administration of Communications in Great Britain to come to an agreement on this with the Ministry of Communications of the U.S.S.R. which from its side supports such a proposal.'

It was a formidable list of requests and instructions and it seemed to me that the direct telegraph link might well be the most difficult. I replied to Blagonravov on 31 January suggesting we begin on 21 February, sent some practical details and then placed the affair in the hands of the Post Office.

Lovell to Taylor Post Office Engineering Dept.: ... the translation of the letter from Blagonravov is given at the end of this message. We suggest that we commence tests on Feb 21st but this will be subject to the Post Office arrangements for direct telegraphic (telex?) communication with Gorki (Zimenki) as specified on page 3 of Blagonravov's letter. May we assume that you will take immediate action on this and that you will also take action on the following:

(1) Provide us with an operator, preferably Russian speaking.
(2) Provide us with the necessary apparatus for producing the signal inputs in tests 3 and 5 of Blagonravov's letter [these were requests for special forms of teletype signals].
(3) Arrange with N.A.S.A. to send us the ephemerides of mutual visibility of Echo II between Jodrell Bank and Gorki for Feb. 21st and succeeding days. This is most urgent because, according to the terms of Blagonravov's letter, we have to provide him and Gorki with the exact schedule 15 days before the tests commence and hence we must do this early next week.

(4) Arrange with N.A.S.A. for the renewal and the extension of the contract which they have with the University of Manchester for the use of the telescope at Jodrell Bank. This is important because the operations with Echo II will make it impossible to use the telescope for any other programmes.

Many of the practical problems were speedily resolved. Soon we had two teleprinters side by side near our control room, one linked to Washington, the other to Gorki. The Russians agreed to receive our messages in English but said they would send theirs in Russian. The ephemerides were supplied from Washington and on 5 February we sent to Blagonravov and Ghetmantshev in Gorki our operating times for the first five days beginning on 21 February on the Echo II transits at 13 47 UT and 23 11 UT. At last on 19 February we received from the G.P.O. authority to transmit: '. . . as you will be using the frequency concerned for short periods only, during a relatively limited space of time and at what is now very short notice we propose as an exceptional measure, to dispense with the formal licence.'

Nothing seemed likely to hinder the initiation of these tests on 21 February. But we had not taken account of the deep political suspicions then existing between the U.S.A. and U.S.S.R. and we had certainly not anticipated the arrival of Colonel X, an emissary from Washington, seconded to Jodrell Bank to supervise the tests. His first act on the morning of 21 February was to forbid us to carry on with the schedule of transmissions to Gorki via the balloon. It is true that a political difficulty had been foreshadowed in a letter which I received from Jaffe dated 7 February. Notwithstanding all the international arrangements already made he wrote: 'We wish to wait for Academician Blagonravov's reply to Dr. Dryden's letter prior to coming to any agreement on a schedule of tests from Jodrell Bank to Zimenki Observatory.' A copy of this letter from Dryden to Blagronravov was enclosed, dated 7 February 1964 it contained the following passage. 'If both sides are to derive maximum benefit from the experience of working together with Echo II the experiments should involve both transmission and reception by each side. I would like, therefore, to ask the Academy of Sciences to give the most serious and urgent consideration to an expansion of the programme of tests you have outlined so that it will provide for similar transmissions from Zimenki to Jodrell Bank.'

Colonel X's instructions were simple—he must prevent us from operating until the Russians had replied giving their agreement to transmit as well as receive. We were annoyed. I said to Colonel X that we were a University establishment and that he had no authority to prevent us turning our telescope onto the balloon. But Colonel X had many contacts and within a short time I was advised from a level which I did not consider it expedient to disobey that we should not carry on with our schedule because 'after all it is an American balloon'. Fortunately the immediate situation was saved, and in my opinion the entire series of tests were rescued from destruction, by J. G. Davies, who in a flash of inspiration said 'but the Moon does not belong to the Americans, we will transmit to Gorki tonight via the Moon'. And that is precisely what we did. A straightforward telex message to Ghetmantshev suggested that it would be advisable to carry out the first contacts via the Moon instead of Echo II. 'To Dr. Ghetmantshev and Colleagues at Zimenki: We at Jodrell Bank take this opportunity of sending our warm greetings to you via the Moon. We hope that this co-operative experiment may lead to closer links in the future between the astronomers and scientists in our countries. With best regards. Lovell, Davies, Thomson, and the staff of Jodrell Bank.'

Opposition to the tests via the balloon then crumbled immediately. There was no evidence, as far as we were aware, that the Russians ever contemplated transmitting as a part of the original international agreement, or that they had any apparatus at Gorki which would enable them to do so. In any case the by-pass of the balloon by using the Moon, clearly made nonsense of further opposition. Colonel X left Jodrell Bank and on 22 February—a day late—Gorki received our signals via Echo II. 'Blagonravov and Ghetmantshev to Lovell, Davies and Thomson: Congratulations on success of first session of experimental radio connection via Echo II satellite as radio wave reflector. Glad emphasize importance of this practical step on international co-operation in field of exploration and utilization of outer space for practical purposes.'

With the Echo II link established, we were then able to transmit to Gorki the following message:

To Academician Blagonravov. This teletype message is transmitted from Washington via the U.S. active repeater Communication Satellite Relay II, to the General Post Office terminal at Goonhilly Downs in Great Britain whence it is retransmitted by the Jodrell Bank Observatory via the U.S. passive reflector communication satellite, Echo II, to Zimenki Observatory in the U.S.S.R. We hope this modest beginning will lead to further association between our countries in international efforts towards development of a global satellite system to serve the communication needs of all nations. We look forward to extending these experiments and to beginning joint activity in the other two programmes of a bilateral space agreement of June 8 1962— the co-ordinated meteorological satellite programme and the magnetic field survey by satellite. With warm regards. Hugh L. Dryden.

The experiments continued without incident until 8 March, by which time all the forms of transmission specified by Blagonravov had been accomplished. These included facsimile picture transmissions. The Russians had no idea what we were transmitting. Over the telex we received the following message: 'Jodrell Bank de Zimenki. We have received the following picture: Some pieces of continuous geocentric circles between them there are four dotted circles. It seems to us that we have seen on these circles the planets signs of the solar system.' They had! J. G. Davies had sketched the planetary system. His original and the picture as received at Zimenki are shown in Plate 5. We were happy that the international astronomical language had survived the vicissitudes of transmissions through space.

A friendly exchange of messages on 8 March marked the end of this experiment apart from some further Moon tests at the end of April which Davies arranged with Ghetmantshev in order to clarify one or two puzzling features of the balloon tests. On 6 March I received an encouraging letter from Garrett, the Scientific Attaché in Moscow. 'The Soviet Press has carried a large number of reports during the past two weeks on the joint experiments carried out by scientists at Jodrell Bank and Zimenki using the American Echo II and the Moon. In every case the reports have been enthusiastic about the success of this international experiment and I enclose English versions of a few articles which have recently appeared on this and related subjects.'

That would have been a pleasant ending to a time-consuming use of our resources and personal energies, which contributed nothing to our real research programmes but which we willingly gave in the interests of international collaboration. Unfortunately real life is rarely like this and the reality in this case was no exception. On 12 June F. J. D. Taylor of the Post Office Engineering Department wrote: 'When I was in the U.S.A. recently Leonard Jaffe of N.A.S.A. mentioned that his Organisation had not then received from the U.S.S.R. any information on the results of receptions at the Gorki Radio Astronomy Laboratory. We too have heard nothing from the U.S.S.R.' In reply I wrote: 'A week or so ago we received a comprehensive document from Gorki setting out the results of the tests, including a magnetic tape. Jaffe has also had a similar communication from them, and he has sent J. G. Davies a translation of the Russian text.' In fact, Ghetmanshev's pleasant covering letter was dated 16 May and with it he conveyed well-bound and comprehensive volumes with many illustrations of their reception of our transmissions. Since we had also provided N.A.S.A. with a long report we could not imagine what more they expected.

Even three years later N.A.S.A. still seemed to harbour some grudge. *2 August 1967 Lovell to Frutkin*: When Graham Smith came back from his recent American visit he mentioned that in his conversations with you he detected that there might be a feeling of uneasiness in N.A.S.A. about the co-operative work which we did on the Echo II programme with the Soviets. This bothers me because I confess to being in a state of blissful ignorance that there were any difficulties other than the purely technical or administrative ones at the time of the experiment, or any lingering feeling of dissatisfaction about the results.

Frutkin merely sent rather a formal reply: 'I think the official situation has been put fairly clearly in the report which our people did on the Soviet/U.K./N.A.S.A. Echo II experiments.' In fact, the failure of N.A.S.A. to persuade the Soviets to transmit as well as receive, and our evasion of the demand not to begin the tests with Gorki until the Soviets had agreed to transmit, coloured the U.S. attitude to these experiments. The most recent official U.S. assessment is contained in the article by J. G. Whelan on 'Soviet Attitude towards international co-operation in space'.* In this he writes with respect to the magnetic field mapping programme for satellites that '. . . this project was never really implemented to any great extent by the Soviet Union. We were disappointed in the results that we obtained there'. He then continues with his summary of the communication experiments.

Similarly, the co-operative communications experiments with the American passive satellite Echo II, using the Jodrell Bank and Zimenki facilities, had mixed results. In 1966, Mr. Webb† reported to Congress on these experiments that had just only been completed. According to Mr. Webb, the Russians observed the 'critical inflation phase of the satellite optically and forwarded the data to us'. They did not forward the desirable radar data, he said, but they had not committed themselves to do so. Moreover, the Russians provided recordings and other data of their reception of the transmissions via Echo from Jodrell Bank. On the other hand, the communications were carried out only in one direction instead of two, and as Mr. Webb said 'at less interesting frequencies than we would have liked, and with some technical limitations at the ground terminals used'.‡

A year later, however, Dr. Seamans made a favourable appraisal of the communications experiment. 'The Soviet Union did gather a great deal of scientific information which,' he said, 'they made available to us'. And he added this note of praise: 'My understanding is that this was done in a very high quality manner by them.'§ An official assessment in 1969 acknowledged that the Soviet Union 'did provide reasonable data relating to their radio receptions via Echo II', but, it added regretfully, 'technical difficulties (partly at Jodrell Bank) limited the experimental results.'¶

* *Soviet Space Programs 1966–70*, Rept. for Committee on Aeronautical and Space Sciences, U.S. Senate, 9 Dec. 1971, Ch. XI.

† Mr. Webb was at that time the Chief Administrator of N.A.S.A.

‡ Senate, Hearings on N.A.S.A. Authorization, FY 1967, p. 38.

§ Senate, Hearings on N.A.S.A. Authorization, FY 1968, pt. 2, p. 938.

¶ U.S./U.S.S.R. Co-operation in Space, 26 Sept. 1969, p. 3. Information provided by the Department of State.

The indication in Mr. Webb's comment and in the official document issued by the Department of State that there were technical difficulties 'partly at Jodrell Bank' astonished those of us at Jodrell Bank when we read these extracts in a volume which came to us in 1972.* The operations proceeded precisely as agreed with the U.S.–U.S.S.R. representatives—and, as for the frequency, this was exactly that specified by the Soviets and, as already described, caused us some difficulty in the reclamation of equipment which had already been abandoned.

Our participation in this Echo II experiment underlined the difficulties which occur when international scientific collaboration is organized on a political basis. It contrasted sharply, for example, with the ease and smoothness of the combined Soviet, U.S., and Jodrell work on the flare stars (Chapter 13). It is doubtful if, in the end, the Echo experiments had any useful scientific or technical consequences. Their major value was the maintenance of contact in the space field between the U.S.A. and the U.S.S.R. at a most difficult period of their relations.

* *Soviet Space Programs 1966–70*, U.S. Senate Report 9 Dec. 1971.

16

The Mark 1A radio telescope

I wrote the first pages of this book in the autumn of 1970 shortly after the Mk I telescope had been placed in the hands of the engineers for modifications to complete the changes converting it to the Mk IA. At that time, particularly since so much of the ground level work had been completed, there was no reason to doubt that the telescope would be working again by the late spring or early summer of 1971 as promised by the engineers and the United Kingdom Atomic Energy Authority (who were acting as agents for the Science Research Council). We settled down to a winter of discontent deprived of our major research tool. But it turned out to be a very long winter, enclosing a spring, a summer, and the beginnings of another winter before we achieved even partial use of the telescope again.

Then, on 14 November 1971 the wheels begun to turn once more. Ironically, it was another Soviet space flight which enabled us to extract some use from the telescope before the engineers had finished. In 1957 it was Sputnik, now it was two probes to the planet Mars—Mars 2 which reached the planet on 17 November and Mars 3 arriving on 2 December. When they reached their target the planet was more than 90 million miles from Earth. From that time forward we had partial use of the telescope, at night and during weekends when the engineers were not at work completing a seemingly endless series of miscellaneous tasks. Finally in mid-July of 1972 the engineers cleared the site and at last we could programme our researches on a twenty-four-hour basis. Now in the autumn of 1972, two years after the first chapter was written, it seems fitting to end this book in the mood of optimism engendered by the success of this conversion.

It would be tedious to detail the story of the conversion. As far as I am concerned the construction of radio telescopes seems to be like a repetitive series of gramophone records: delays and financial problems calculated to drive the astronomer to despair and then, in the end, success beyond expectation. Although it is not my intention to describe the daily details of these familiar problems, it is important to the story of the development of radio astronomy at Jodrell Bank—and indeed in the U.K.—to give some account of the strategy which led to the decision to spend over £600 000 (as much as the Mk I cost originally) on this process of renovation and modification.

The indication that I must have had many informal discussions with Husband and with D.S.I.R. about improving the Mk I, quite soon after it came into use, is contained in the first formal approach which I made to D.S.I.R. in 1963. On 23 January the University made application on my behalf for a grant of £4000 for the 'Investigation of deflections and ancillary problems of the Mk I radio telescope'. In the full account of the investigation I wrote:

The Mk I 250 ft radio telescope at Jodrell Bank has been in operation since the autumn of 1957 and since that time it has carried out over 20 000 hours of research work on radio astronomical problems covering a range of frequencies from a few megacycles to the hydrogen line frequency of 1420 Mc/s. When the telescope was constructed there were no techniques for radio astronomical investigations on high frequencies, and the main emphasis on the Mk I was to obtain a large aperture and power gain in the lower regions of the frequency spectrum. During the course of construction certain modifications were carried out to the telescope as a result of the discovery of the hydrogen line emission on 1420 Mc/s., and it was hoped that the telescope would be effective with an aperture of about 100 ft on this frequency. In fact the performance of the telescope has far exceeded this expectation. . . . During the last few years developments in low noise amplifiers at very high frequencies (for example, masers and parametric amplifiers) have opened new fields for radio astronomical studies in the frequency bands extending up to several thousands of megacycles. The concept that it should be possible to modify the Mk I telescope to work in these higher frequency bands and at the same time to attain the theoretical performance of the 250 ft aperture on the hydrogen line frequency has often been discussed with D.S.I.R. and it appears in various documents as the Mk IA proposal. However, it has always been considered that it would not be desirable to take any steps along these lines which would put the Mk I telescope out of operation for a considerable time until the Mk II facility became available. There is, in fact, every indication that the Mk II telescope will be operational towards the end of 1963 or at the latest early in 1964,* and it is now desired to take the preliminary steps towards the Mk IA concept. The appropriate procedure was discussed between Messrs. Husband & Co. and the Ministry of Works at Jodrell Bank on January 25th, and it was agreed to be essential to first of all make an accurate measurement of the deflections of the present telescope under various operational conditions (particularly the change in shape of the reflector with elevation of the bowl). . . . As a result of this investigation it would be hoped that an estimate of the feasibility and cost of making various degrees of improvement in the shape of the bowl would be available. From the astronomical aspect it can be said

* See Ch. 7; tests on the Mk II telescope began in the summer of 1964.

straight away that an improvement even by a factor of 2 would be of extreme value, since this would lead to the achievement of the theoretical performance appropriate to the 250 ft aperture on the hydrogen line frequency and would give an equivalent aperture of the order of 200 ft on 10 cm.

This application went through D.S.I.R. without undue trouble, and in mid-August of 1963 I was able to let Husband know that Treasury approval had been obtained.* The formal notice of the award of the grant was made to the University on 3 December 1963. 'In awarding this grant the Department of Scientific and Industrial Research is not committed to financing any improvements or modifications to the Mk I telescope proposed as a result of this or any other investigation.'

Earlier in January Husband wrote to me:

... until the investigation has been made, which will take quite a long time because it will mean taking accurate measurements as the bowl rotates, we cannot say what greater accuracy it may be possible to obtain or the best method of going about the job. You mentioned that it would probably be worth while carrying out major improvements if the accuracy of shape could be improved by a factor of 2 and any modifications which would steady the elevation movement when the telescope is working on a very narrow beam would be extremely valuable. I have in mind a supplementary drive, or supplementary drives, through steadying rollers, probably on two slightly smaller bicycle wheels. The main drive would be maintained as at present but a relatively small additional horsepower could be applied through the friction of the rollers to adjust overall torque.

On 19 August 1963 I wrote to Husband:

In connection with the Mk IA ... there is no reason at all for any further delay on this matter other than that occasioned by limitations on the availability of the telescope. . . . I am anxious to do everything I can to facilitate this Mk IA investigation because I think it will have to be the next big thing which we push for. The multi-million pound Mk V project will certainly come, but both you and I are only too well acquainted with the delays which always occur, and I am quite certain that the Mk IA modification should be the next major job, with an aim that we might perhaps commence a few months after the introduction of the Mk II into our programme.

Husband quickly completed the investigation of the deflections of the Mk I. Early in January he wrote to me to convey the major results of the investigation and to give an outline of his proposals, and in April 1964

* Because of the previous financial history of the Mk I, D.S.I.R. deemed it desirable to obtain Treasury approval although in principle there was no statutory obligation to do so for this small sum. The point at issue, of course, of was that the investigation might well lead to a future request for a sum in excess of £100 000 to be spent on the Mk I.

we had in our hands his report: 'A survey of the reflector and proposed modifications to the Mk I—250 ft radio telescope.'

The timescale of the proposed changes, and my own reluctance to put the Mk I out of action, soon led to a situation where there was a conflict in principle, if not in money, between the proposal for the 400 ft diameter Mk V and this Mk IA conversion. In fact, the situation had already developed where I was pressing forward with the Mk V concept, which arose from the ashes of the Mk IV idea described in Chapter 7. In the month of April 1964, when we received Husband's report on the conversion of the Mk I, the University courageously entered into an agreement with Husband to make a feasibility study for a steerable telescope 'not less than 375 ft (major axis) by 250 ft (minor axis) and a maximum size of 500 ft by 250 ft'. The concept became known as the Mk V, and although our negotiations were carried out with the full knowledge and approval of D.S.I.R. they would not provide the money for the investigation. At that time the extraordinary situation existed that D.S.I.R. were not allowed to spend money on preliminary investigations of scientific projects unless they could give the necessary assurance to Treasury that it was their intention to proceed with the project, and that the money was available in their allocation. In February of 1965 we had the draft of Husband's report on 'Proposals and cost estimates for the Mk V radio telescope'. The report was accompanied by a curve giving the costs for a telescope of circular aperture ranging from £1·785 millions for an aperture of 260 ft to £4·085 millions for a 400 ft aperture (at 1965 prices). The report was sanguine as regards the possibility of constructing the 400 ft telescope, but for larger instruments the curve of cost against size rose so steeply that the 400 ft became our ultimate aim.

If the Mk V is ever built I hope that someone will describe in detail the maze of scientific and administrative arguments which sterilized progress from that moment in 1965 when we had Husband's report before us. Clearly I could not press simultaneously for this telescope and also for the significant sum of money required to modify the Mk I. Not the least of my personal difficulties was that I was appointed Chairman of the Astronomy Space and Radio Board on the formation of the Science Research Council in 1965. It was not possible to perform the duties of a Chairman in any satisfactory manner and simultaneously fight as hard as I otherwise would have done for such a large share of the Board's available money.

However, by the end of 1967 two separate factors impelled me to take a decisive stand for the Mk IA at the expense of a conscious decision to defer the Mk V for a year or so. The first of these was internal to S.R.C.

and especially to the affairs of the Astronomy Space and Radio Board. This board had, on its programme, and in its 'forward look' a number of expensive and important items. Many of these were in the multi-million-pound category—for example, the Anglo–Australian 150 inch optical telescope (then already under construction), the series of U.K. satellites—as well as heavy demands for routine expenditure and minor developments for the Royal Observatories, the Radio and Space Research Station and various University departments. In the sphere of radio astronomy the two major items lying in the future were our own Mk V proposal, and Sir Martin Ryle's proposal for a 5 km aperture synthesis instrument at Cambridge. In the autumn of 1967 the estimates for these two were in the region of £4 to £5 million and £2 to £3 million respectively (at 1967 prices). Taking account of the timetable for construction and the incidence of expenditure in the ensuing years it still seemed possible to accommodate both of these projects simultaneously in the Board's budget.

At that moment, however, the Science Research Council received additional and severe cuts in the growth rate which it had so far been able to maintain. Even with the priorities which the other interests in SRC were prepared to give to astronomy, it was evident that the Board could no longer proceed simultaneously with the Cambridge and Jodrell concepts. At the end of the December 1967 meeting of the board the Chairman of the Council, Sir Brian Flowers, asked to see Ryle and me. His message was not unexpected. A choice had to be made. Did Ryle and I wish him to 'toss a coin'? Throughout the years the strength of radio astronomy in the U.K. had depended on the mutual understanding between Ryle and myself and I had no desire to place this in jeopardy at that critical moment. In the event I had no hesitation in replying that Ryle's telescope should be built first. He was ready, there were still scientific arguments about the Mk V and the final design was not done. Furthermore, his instrument was only half of the estimated cost of the Mk V and was in international competition to a greater extent than the larger dish. I made one condition—that priority would be given by the Board to the conversion of the Mk I to the Mk IA at a cost then estimated to be less than £½ million.

The second factor related to the condition of the Mk I telescope. Both in *The Story of Jodrell Bank* and in this book I have emphasized the extraordinary continuity of work which we were able to achieve with this instrument. But even the excellent maintenance of Commander Tolson and his group of engineers could not prevent some deterioration in the foundations and the structure. Husband gave me many warnings and in 1966 they became more urgent.

Husband to Lovell 10 June 1966: Whilst appreciating the immense importance of keeping this telescope in commission until you have another big one we must not overlook the fact that we reported on certain weaknesses about three years ago, and the desirability of providing further support to control the deflections on the bowl structure—which would automatically relieve the present turntable of some of its load. If the Science Research Council can give no indication as to when a real start can be made on the second big aerial then I must advise that it would be wise to draw up a programme for improving the Mk I at an early date. By next year the telescope will have been running for 10 years, which is a very long time for what was virtually a prototype.

On 13 June I replied: 'We are all impressed with the importance of doing whatever is necessary to keep the Mk I in service and of the urgency of making a real start on the Mk V.... The present position on the Mk V is that after a long period of gloom I now think that there is a little hope ... that we should be able to make some active progress with this instrument.'

Early in 1967 there was a warning about the condition of the azimuth rail track.

Husband to Lovell 11 February 1967: I understood from Tolson that four more anchor bolts have broken. ... Where the cement packing has deteriorated under the rail base plates, I think by frost action over the years, this should be replaced under very close supervision at the first opportunity of putting the telescope out of commission. ... Our recommendation still stands that the telescope as a whole should be modified by building an inner ring in order to improve the efficiency and to relieve the track and foundation loading.

In my reply of 15 February I remarked that I had '... constantly in mind your recommendations concerning the Mk IA modifications. As I explained to you on Friday our decisions as to the extent to which we proceed with this depends very much on the outcome of the Mk V design studies, and as these progress I hope we will be able to form our opinion on the Mk IA issue.'

In the following weeks there was much further discussion between Husband, Tolson, and myself about the condition of the track, and in mid-March I wrote a long letter to the Bursar, Rainford, to let him know that we were obtaining estimates from Thornton's for attention to some of the bogies and from T. W. Ward's for a test re-lay of a section of the track with the suggestion that 'the University's insurers would regard this as a case of fair wear and tear and the cost of the remedial work should be borne by them'. Rainford was quick to reply that he 'did not see that the cost of the remedial work ... would be accepted by our insurers. As you will no doubt remember the railway track is specifically excluded from cover.' In the end, he authorized the expenditure of a small sum of money (about £800) necessary for Wards to re-lay a part of the track during that summer, and the S.R.C., although emphasizing that they could 'not accept any general obligation to maintain an instrument long since handed over', nevertheless agreed to consider an application for a special grant of £4200 for the repairs and modifications to the bogies. Rainford asked that I should also cover the track modifications in this application. On 24 May 1967 my application to S.R.C. requesting £4200 for modification of some of the pins in the azimuth bogies also contained the following explanation about the state of the railway track.

The concrete packing under the track started breaking up in certain places in the winter of 1962. Repairs were carried out in the summer of 1963. Since that time the concrete packing has continued to break up and the rate of deterioration has increased causing considerable concern. Husband & Co. advised the taking up and re-laying of a sample section of the track for examination, to try to establish the reason for the breaking up of the concrete packing, and to determine in due course the best method and means of repairs of the track as a whole. The sample section of track has been lifted and re-laid, the estimated cost of the work being £775. The actual final cost of this work is not yet known.*

Unfortunately a situation far more serious than that indicated by the condition of the bogies or the azimuth track was revealed in September of 1967. During a routine inspection one of the engineers noticed a thin line of rust on one of the massive steel cones which carry the whole load of the bowl to the trunnion bearings at the top of the towers. Removal of the rust revealed a thin hair-line crack, and others were found in the other cone. On 19 September Handyside† received a letter from Kington‡. 'Confirming my telephone conversation with you today, September 18, the cracked welding on the elevation drive cones shows a crystalline surface indicating that the failure is most probably due to fatigue arising from the stressing cycle over the past ten years of the telescope's use. . . . We feel it would be wise to limit the maximum speed in elevation to not more than 5° per minute and to avoid all sharp accelerations.' A few days later I received a long letter from Husband. 'On Friday, the 15th, Richard Husband§ . . . climbed into the cones and

---

* The S.R.C. considered, quite properly, that a retrospective payment in this way was not in order. Fortunately, in the end, the cost of the work (£553) was less than the estimate.

† Commander D. Handyside, our Deputy Chief Engineer at Jodrell Bank.

‡ C. N. Kington, M.B.E., a senior partner of Dr. Husband.

§ Dr. Husband's elder son who had now joined the firm of Husband & Co.

closely inspected the fractured welds. . . . There is little doubt that the cracks are due to fatigue effects caused by the alternating stresses set up when the bowl is driven in either direction.' The letter then gave an explanation of the structure of the cones and continued:

Unfortunately, although the local cracking of the welded connection has caused a re-distribution of the torsional load, and to a much lesser degree the gravitational and wind loads, this cannot be accepted as a permanent state of affairs. Once a crack has started in a fillet weld, and the remainder of the weld remains in use, the crack tends to spread. In a period of time which I cannot estimate, the racks would eventually become entirely disconnected from the cones and would only be held to the bowl structure by the secondary connections referred to above. If the telescope was still allowed to remain in use these secondary connections, which would then be overloaded in torsion, would fatigue and fail and the rack would fall out of mesh, probably causing tooth damage and perhaps jamming the bowl against any further rotation.

Husband then continued with further emphasis on the safety precautions which we were to take during operations, and a list of the procedures for temporary remedial action which he was negotiating with United Steel. The letter ended:

Finally may I remind you that the useful life of the Mk I can only be prolonged for a considerable number of years by putting in hand the more extensive improvements put forward by us in our report of April 1964. In that report we recommended relieving the trunnions of a substantial part of their present loading; this in turn would automatically increase the capacity of the structural system to survive the combined torsional, gravitational and wind loads for an indefinite period.

At last I was thoroughly frightened about the condition of the Mk I telescope. For the first time I began to realize that it might be impossible to have a Mk V telescope in operation before the Mk I became unusable.

Lovell to Rainford 31 October 1967: I feel I ought to express to you my anxiety about the Mk I telescope. It is convenient to deal with this under two headings.
1. First of all I want to express my concern about the cost of the welding repairs to the fatigue cracks in the cones supporting the bowl. After the first inspection Husband and Hewson of United Steel said that the repairs would be cheap and would be unlikely to exceed £500. . . . On Oct. 16 . . . a sum of £2570 was mentioned. I will be surprised if the final bill is not greater than this. . . . we have been plagued by two factors. First of all the long run of windy weather has inhibited continuous work which has meant that for a considerable amount of time the men have been idle here. Much more serious is the fact that as soon as some of the cracks were welded then others were discovered, and worst of all new cracks developed. . . . the fact remains that the

net result of several weeks work is that the repairs to one tower cone are not yet complete. This tower, is in fact, today being abandoned temporarily and the other one tackled. I strongly recommend you . . . to negotiate with the insurance Company. . . .*

2. These troubles have drawn my attention most forcibly to the warnings which Husband has repeatedly given us that we would have to consider relieving the load and stresses on parts of the Mk I if we were going to keep it in operation. You will recall that in 1964 we asked Husband to present us with a report for the modification of the Mk I telescope to the Mk IA concept. At that time we had in mind the replacement of the membrane, the introduction of a quadrupod as well as the necessary mechanical alterations to relieve the stresses. Husband's estimate at that time was £268 400. However a large amount of this work was concerned with the improvements to the radio performance of the telescope, and certain works concerning the bogies and railway track have already been carried out. The estimate in their report for the modification of the steelwork was £200 000 *but* this included the introduction of a new membrane and quadrupod instead of the aerial tower. I have now asked Husband to let me have his current estimates of the cost of the part of these repairs necessary only to relieve the mechanical stresses, that is the introduction of the modification to the bicycle wheel and the inner track without any subsidiary changes to the instrument to improve its radio performance (that is excluding the new membrane and the quadrupod). . . .

Clearly I was still hoping that the Mk I could be saved mechanically at a cost which would not endanger the Mk V negotiations. But it was not to be, since on 10 November Husband reminded me that it would be impossible to carry out part only of the Mk IA concept as I had suggested. 'The main difficulty in carrying out part only of the proposed modification put forward in my 1964 report is that the additional bicycle wheels were intended to be balanced by the weight of the new reflector and quadrupod. We could consider fitting the bicycle wheels and balancing them with temporary balance weights at their upper extremities. Unfortunately, these would add undesirable concentrated loads where we least desire them.' Although this letter removed my hopes of a quick and cheap solution to the problems it also contained more cheerful news about the cracks.

I am now firmly of the opinion that these cracks have existed for a long time. They are due to eight angle cleats, forming part of the connection between each cone to the main bowl structure, acting in a manner which was not anticipated in the original design. The proper purposes of the cleats, for which they were adequate and important, is to assist in the transfer of turning moments from the bowl structure to the driving racks and vice versa. Un-

---

* Who refused to entertain a claim. Eventually after S.R.C. had approved the Mk IA conversion, they also agreed to make a separate grant to cover the cost of this repair work on the cones.

fortunately, the configuration adopted tends to cause stretching of the rivets attaching the cleats to the cone face plates in addition to transferring torques via their shear resistance. . . . We therefore have a situation in which the cleats, the rivets and a section of the weld *immediately behind the cleats* is subjected to a considerable variation of stress at each elevation motion of the reflector.

Husband again emphasized that the useful life of the telescope could be prolonged for an indefinite period only by carrying out the proposals of his 1964 report.

In my reply of 27 November I wrote: 'One thing that is absolutely certain at the present time is the utmost importance of securing the future existence of the Mk I telescope, and I am anxious to have in mind the possibilities in this respect either to the extent of a complete Mk IA modification or a minor version of it to deal with the mechanical problems only.'

This was the acutely urgent operational background which led me to take the decisive stand for the Mk IA modifications in the meeting with Sir Brian Flowers and Sir Martin Ryle a few weeks later. Subsequent events have confirmed the critical nature of that decision. Fortunately the proposal to proceed with this, and with Ryle's new instrument, whilst delaying the Mk V for a year or so, made such good scientific and economic sense that the proposals were accepted with alacrity in all quarters. I acquainted the Vice-Chancellor and Rainford with the new proposals and on 3 February 1968 they made formal application to S.R.C. on my behalf for a sum of £350 000 to carry out this work. During the March and April meetings of the respective Committee, Board, and Council in S.R.C. the proposals went through smoothly, and since the Secretary of State for Education and Science had been present at a meeting of Council when the astronomical programmes, including this concept, was discussed* we did not anticipate any difficulty in obtaining the approval of the Treasury. Indeed the formalities were soon completed and on 8 July 1968, the S.R.C. was able to make a public announcement of

a grant of £400 000† . . . to meet the cost of repairs and engineering modifications . . . which will extend the working life of the telescope and will enable the present research programmes to be continued and expanded. The work is expected to take up to two years. . . . it is expected that the proposed repairs and modifications will also result in a significant improvement in the performance of the telescope at shorter wavelengths, yielding full theoretical

* See Ch. 10.
† This exceeded the grant of £350 000 requested by the necessary amount to cover engineers' and agents' fees together with some contingency.

efficiency in the 18–21 cm range and efficient operation at still lower wavelengths in the central portion of the bowl.

The essential features of Husband's 1964 proposals were to build a
new reflecting surface above the existing membrane. The existing stabilizing girder was not load bearing and it was therefore necessary to devise a means of carrying the additional load of this structure as well as
relieving the trunnion bearings and the cone structure of some of the
existing load. It was proposed to build two new load-bearing wheel
girders placed 48 ft each side of the stabilizing girder. These would run
on upthrust units situated above the central part of the diametral girder,
and the load would be transferred to ground by a structure enveloping
this central section of the diametral girder, carried on undriven bogies,
running on a new inner azimuth track.

In fact, this is precisely what was eventually done over six years later,
although there were naturally some changes in detail. The most significant of these changes concerned the arrangement of the aerial tower.
In the 1964 report Husband proposed that the aerial should no longer
be carried on the single tower; '. . . the new smaller tolerances can only
be met by spreading the base of the aerial supporting structure very
considerably and it is for this reason that we recommend a replacement
of the existing mast by a tetrapod with legs spread approximately 120 ft
diagonally at the points of attachment to the main structure.' The focus
was intended to remain in the plane of the aperture.

The decision to make a change from this aerial arrangement proposed
in 1964 was taken because of the great importance assumed by polarization measurements in the intervening years. For these it is essential to
have complete symmetry in the aerial support arrangements. In February and March of 1968 we had many consultations with Husband
about the scientific and mechanical consequences. The aerial support legs
proposed in the 1964 report were asymmetrical and in a memorandum
dated 19 March, dealing with several points of detail I wrote:

There has been a further realization of the importance of polarization
measurements since the original Mk IA proposals were drawn up in 1964, and
in the last few years the Mk I telescope in its present form has undertaken a
series of important polarization measurements of the background radiation
and of the polarization of sources (the latter in collaboration with the Mk II).
It has therefore been necessary to give careful consideration to the preser-
vation of symmetry in the aerial feed arrangements in the Mk IA modifications. For mechanical reasons the original scheme . . . proposed an asymmetrical quadrupod which would not be satisfactory for polarization measurements. The alternatives discussed with the engineers have shown only two
real possibilities to maintain the polarization performance. One is to retain a

quadrupod design but to make it symmetrical, which would imply that the legs would have to be on a circle of approximately 60 ft radius, or to retain the existing tower. We have considered this situation with great care and have decided to retain the central tower arrangement.

Unfortunately with this modification Husband found that the structure would be no longer in balance—the centre of gravity of the bowl structure had to be retained along the trunnion axis—and with the removal of the quadrupod the system became seriously out of balance. During a small party which the University gave a few days after the publication of *The Story of Jodrell Bank* in late May, Husband gave me a rough sketch conveying his solution. As far as we were concerned it was most agreeable—he proposed to restore the balance by *increasing* the size of the new membrane from 250 ft to 265 ft diameter with a consequential increase in focal length to 82·15 ft (an f/d ratio of 0·31).

Husband's final report with cost estimates of these proposals for the conversion of the Mk I to IA was dated July 1968. The report contained the analysis of the weight distribution. To relieve the existing cone and trunnions a load relief of at least 200 tons was projected. The new membrane, framework, and bicycle wheel structure would add 723 tons (membrane 220 tons, new supporting steelwork 148 tons, connections to old framework 55 tons, new 'bicycle' wheels 300 tons). An upthrust of 960 tons was proposed, giving an eventual load relief of 237 tons on the trunnions. The current cost estimate had risen to £374 250 and it was proposed to construct the new inner railway track (Phase 1) during the late summer and autumn of that year 1968 (during which we would have limited azimuth but full elevation motion), and then to commence Phase 2, the major engineering work on the bowl, on 1 April 1969 with an estimated completion date by 31 December, during which time we would have no use of the telescope for research.

In the summer of 1968 it seemed that all would go smoothly. With the U.K.A.E.A. as agents, the S.R.C. had full and detailed knowledge of these various discussions and changes. The first meeting of the 'Mk IA Project Committee' was held at Jodrell Bank on 19 July. J. Hosie of S.R.C. was Chairman*. He complained bitterly about the new estimate of £374 250, which had to be compared with the £350 000 in view when the grant of £400 000 was made only a few weeks earlier. He said that the authority which S.R.C. had from the Treasury would certainly not extend beyond a 10 per cent contingency, which meant that the whole project including the engineers' and agents' fees had to be contained

* J. Hosie, Director of the Astronomy, Space and Radio Board Division; that is the permanent S.R.C. office parallel to the Independent Chairman of the Board, a post which was held by me until September 1970.

within £440 000. The agents and Husband said that the cost figure depended heavily on the estimates for the steelwork, which was a complex constructional job. Husband said that he had taken a current figure of £250 per ton, but it was impossible to refine this figure until actual tenders were obtained. There was only one course possible, and that was the obvious one of asking Husband to proceed with the detailed drawings in order to get competitive tenders. In the meantime it was agreed to continue with the construction of the new inner railway track. At the first of the site meetings on 5 September it was agreed to award the contract for the piling to the same firm who had constructed the original Mk I track—Cementation Ltd. So, in the last week of September 1968 we once more heard the sound of pile-driving on the site of the Mk I—almost exactly sixteen years after the original piling began for the azimuth track of the instrument. This work proceeded without particular incident, or indeed without serious interruption to our use of the telescope, apart from restrictions on azimuth movement during the working day, and it was effectively completed early in 1969.

But the scheme to proceed with Phase 2 (the major steelwork and membrane modifications) and complete the work between April and December 1969 never materialized. On 13 December 1968 I wrote to Rainford a letter of an all too familiar type. 'I am sorry to have to inform you that chaos has developed on the Mk IA grant. At the site meeting yesterday morning we were presented with the tenders for the steelwork. ... the estimate for the steelwork was £142 750. The lowest tender received was £248 000* from Teesside Bridge. ... U.K.A.E.A. estimates that allowing a 10 per cent contingency, the bill is now £504 000 excluding fees against the £350 000 excluding fees ... which we have in our budget.' I was reluctant to go back to S.R.C. to ask for more money for this project; to do so would I feared put an end to any hopes of the Mk V. With this sentiment Hosie entirely agreed.

There followed weeks of discussion during which the whole concept of the change was examined again with an insistence by me that a search should be made for the cheapest way of achieving the essential load relief on the trunnions. Eventually on 20 January I acquainted Rainford with the outcome.

... the present position is that Teesside Bridge are submitting new tenders for another version of the Mk IA which is, in fact Husband's original scheme with the aperture remaining the same, that is 250 ft instead of the 265 ft which he proposed last spring. ... with some sacrifices he can reduce the weights to achieve the necessary balances, and there seems to be a general confidence

---

* The actual tender figure of £364 352 included the membrane. This compared with the estimate of £258 750.

that the cost of this will be well within the financial ceiling which we have available.

On 7 February 1969 Husband wrote to me: '. . . our recent negotiations with the contractors for the modifications on the basis of the less accurate reflector have resulted in the net saving of £83 724. Again it is our misfortune to be tackling this particular job during a period of rapidly rising prices, much more so than seems to be generally admitted.'

The sacrifices which had to be made were unfortunate, since they involved not only the reduction in size of the reflector, but a change in its contruction. The panels of the original reflector were welded to form a continuous surface Now in the Mk IA Husband proposed to use 14 gauge (0·080 inches) mild steel plate, and to form the membrane from 448 panel groups. Each panel group was to consist of 4, 5, or 6 panels bolted together and supported as a unit by up to 12 screwed adjusters connecting the panel group to the supporting steelwork beneath. Also auxiliary surface shape adjusters were to be provided, consisting of screwed connections between the surface and the frame. This comprehensive scheme had to be abandoned. Weight and cost were saved both by reducing the size to 250 ft and by using larger panel groups so that far more of the membrane could be prefabricated on the ground.* This implied that the surface of the reflector could not be adjusted with the precision which had been anticipated. To compensate for this saving of weight above the centre of gravity, the original wheel stabilizing girder was removed.

These were months of misfortune leading to savings which were so quickly overtaken by increasing costs that, in the end, I had to go back to S.R.C. for a supplementary grant. All hope of modifying the telescope in 1969 vanished. At last it was agreed to proceed with a Stage 2 programme in 1969—the installation of the four new bogies on the inner railway track, and the erection of the structural steelwork tower carried on those bogies. It was also agreed to carry on with the work on the rehabilitation of the outer track and the existing bogies. This proceeded satisfactorily in the autumn of 1969. Meanwhile in July we received news that the Department of Education and Science and the Treasury had agreed to make the supplementary grant of £145 000.

There is little more to add to this story of the processes of thought and action which led to the construction of the Mk IA in its present form. The urgency of the work was again emphasized when cracks were found in December 1969 in some sections of the main framework carry-

---

* 336 panels of 12 gauge mild steel plate, instead of the 448 of 14 gauge originally intended.

ing the bowl to the trunnions. Until the telescope was placed in the hands of the engineers for the final Stage 3 modification in August 1970, we limped through our remaining research programmes with penalizing wind restrictions and elevation movements limited to half the normal speed. The reason for Husband's repeated warnings over the previous years was evidently materializing in a dangerous form. The nine months projected for the work extended to fifteen—it was 14 November 1971 before we were able to claim some use of the telescope once more. The addition of hundreds of tons of new steel and its integration with an old structure at high level was a complex job for the contractors—Teesside Bridge. When I repeatedly complained about the delays, Husband told me that it was the most difficult engineering job he had ever tackled. But at last, after the tensions and anxieties, came the joy of having the telescope working again; a superior instrument, more handsome in appearance, and with an accuracy of surface and precision of control which were beyond contemplation at the beginning of my story twenty years ago.

ABBREVIATIONS AND SYMBOLS

1. Journals

A. Rev. Astr. Astrophys.	*Annual Review of Astronomy and Astrophysics* (Palo Alto, Cal.)
Astr. J.	*Astronomical Journal* (Yale University Observatory, New Haven, Conn.)
Astr. Zu.	*Astronomiceskij Zurnal* (Moscow)
Astrophys. J.	*Astrophysical Journal* (Chicago)
Mon. Not. R. astr. Soc.	*Monthly Notices of the Royal Astronomical Society* (London)
Nature	*Nature* (London)
Nature, Phys. Sci.	*Nature, Physical Science* (London)
Observatory	*The Observatory* (London)
Phil. Mag.	*Philosophical Magazine* (London)
Proc. nat. Acad. Sci. Am.	*Proceedings of the National Academy of Sciences of the United States of America* (Washington)
Proc. R. Soc. London Ser. A.	*Proceedings of the Royal Society of London* (London) Series A—Mathematical and physical sciences
Publ. astr. Soc. Japan	*Publications of the Astronomical Society of Japan* (Tokyo)
Publ. astr. Soc. Pacific	*Publications of the Astronomical Society of the Pacific* (San Francisco)
Q.J.R. astr. Soc.	*Quarterly Journal of the Royal Astronomical Society* (London)
Science	*Science* (Washington; New York; Lancaster)
Sky Telesc.	*Sky and Telescope* (Cambridge, Mass.)

2. Organizations

B.M.E.W.S.	Ballistic missile early warning system
COSPAR	Committee on Space Research. Committee of the International Council of Scientific Unions operating on an international basis for space activities in the same manner as the International Astronomical Union does for astronomy, and U.R.S.I. for Scientific Radio
C.S.I.R.O.	Commonwealth Scientific and Industrial Research Organisation, Sydney, N.S.W., Australia
D.S.I.R.	The Department of Scientific and Industrial Research disbanded in 1965 when the newly formed Science Research Council (S.R.C.) assumed the D.S.I.R. responsibilities for the direct Government support of University researches, see footnote pp. 124, 125.

I.A.U.	The International Astronomical Union
I.C.I.	(in connection with fellowships) The Imperial Chemical Industries
J.P.L.	Jet Propulsion Laboratories (California)
M.I.T.	Massachusetts Institute of Technology
M.P.B.W.	The Ministry of Public Building and Works (U.K.)
N.A.S.A.	The National Aeronautics and Space Administration (U.S.A.)
N.R.A.O.	The National Radio Astronomy Observatory at Green Bank, W. Virginia, U.S.A.
O.N.R.	The Office of Naval Research (U.S.A.)
R.R.E.	The Royal Radar Establishment, Malvern, Worcs, the successor of the Government research establishment which developed radar aids for the Royal Air Force during World War II. It was then successively A.M.R.E. (Air Ministry Research Establishment); M.A.P.R.E. (Ministry of Aircraft Production Research Establishment); T.R.E. (Telecommunications Research Establishment) and R.R.E. (Radar Research Establishment). The establishment moved to Malvern in 1942. Defford was the establishment's wartime aerodrome where the telescopes referred to in this book were subsequently built on the runways.
R.G.C.	Research Grants Committee (of the former D.S.I.R.)
S.R.C.	The Science Research Council, established in 1965, is responsible to the Department of Education and Science for the allocation of Government money for civil research in physics, chemistry, astronomy, engineering, and space in Universities and establishments responsible directly to the Council (such as the Royal Greenwich Observatory)
S.T.L.	Space Technology Laboratories (U.S.A.)
U.G.C.	The University Grants Committee
U.K.A.E.A.	United Kingdom Atomic Energy Authority (used in connection with the agency for construction of the radio telescopes)
U.R.S.I.	Union Radio-Scientifique Internationale (The International Scientific Radio Union)

3. *Other abbreviations and symbols*

Telescope nomenclature

Mark I or Mk I '	the original 250 ft aperture steerable radio telescope at Jodrell Bank
,, Ia	the modified version of Mk I
,, II	the elliptical 123 ft × 83 ft 4 in radio telescope at Jodrell Bank
,, III	the elliptical 123 ft × 83 ft 4 in transportable radio telescope built at Wardle, near Nantwich, Cheshire
,, IV	an extremely large steerable radio telescope which never progressed beyond the feasibility study stage

Mark V — the 400 ft steerable radio telescope, abandoned in favour of Mk VA

„ VA — the 375 ft steerable radio telescope for which the detailed designs were completed in 1973

Source lists

3C 295 etc. — radio source No. 295 in the 3rd Cambridge catalogue of radio sources

PKS 0531+19 etc. — a radio source in the Parkes (NSW) catalogue. The figures refer to the right ascension 0531 (= 05^h 31^m) and declination (+19 deg)

CTA 101 etc. — radio source No. 101 in the A catalogue of the California Institute of Technology

PSR 1749−28 etc. — the nomenclature used for pulsars. The figures give the right ascension (e.g. 1749 = 17^h 49^m) and declination (−28 deg)

Variable stars

EV Lac, UV Ceti etc. — The nomenclature used for variable stars. Normally the individual stars in a constellation are designated by letters of the Greek alphabet followed by the Latin name of the constellation, generally in order of decreasing brightness. For example, α-Gemini is the brightest star in the constellation of Gemini, and β-Gemini the second brightest. In the case of variable stars the international convention is that the name of the constellation is now preceded by capital Roman letters, commencing with R, S to Z, then RR, RS to RZ; until ZZ is reached then the nomenclature AA to AZ, BB to BZ etc. is used. Since the letter J is not used 334 variable stars can be designated in this way. The designation V followed by a number is then used. In the case of UV Ceti the number L726−8AB is also appropriate because it is a double star and double stars are sometimes specially designated by the initial of the discoverer plus the number of the star in his catalogue.

Frequency

Hz hertz
kHz kilohertz (10^3 Hz)
MHz megahertz (10^6 Hz)
GHz gigahertz (10^9 Hz)

used for frequency and by international agreement replaced cycles per second, kilocycles, and megacycles (cs^{-1}, kcs^{-1}, Mcs^{-1}) during the period covered by this book. The old symbols have been retained only where appropriate in quoted passages.

Length, time and astronomical distances

s second

ms millisecond (10^{-3} s)

μs microsecond (10^{-6} s)

ns nanosecond (10^{-9} s)

cm, m, km   centimetre, metre, kilometre,
             but m as superscript (e.g. 10^m) refers to magnitude (of a star
             or galaxy).

pc   parsec $19 \cdot 28 \times 10^{12}$ miles ($30 \cdot 86 \times 10^{12}$ km) unit of dis-
                tance used in astronomy. A parsec would be the

kpc   kiloparsec 10^3 pc   distance of a star having a parallax of 1 second of
                               arc—i.e. the angle subtended at the star by the

Mpc   megaparsec 10^6 pc   radius of the Earth's orbit. 1 parsec = $3 \cdot 26$ light
                               years. In fact the nearest star has a parallax of
                               $0 \cdot 765$ seconds of arc, that is a distance of $1 \cdot 3$ pc or
                               $4 \cdot 25$ light years.
                               (Note that *solar* parallax is the angle subtended at
                               the Sun by the radius of the *Earth*, not the Earth's
                               orbit.)

UT   Universal Time, that is, the mean time for the meridian of Greenwich
      starting at midnight. GMT (Greenwich Mean Time) was commonly
      used in the U.K., and GCT (Greenwich Civil Time) etc. elsewhere
      until the International Astronomical Union adopted the UT nomen-
      clature in 1935.

Miscellaneous symbols, units and constants

H km s^{-1} Mpc^{-1} Hubble constant in kilometres per second per megaparsec.

c velocity of light ($2 \cdot 99 \times 10^{10}$ cm s^{-1})

z redshift (of a distant galaxy). The ratio of the change in wavelength of
       a spectral line ($\delta\lambda$ or $\Delta\delta$) to the actual wavelength (λ).
       $z = \delta\lambda/\lambda$. This equals the ratio of the velocity of recession (v) to the
       velocity of light (c) for small v. (For the relativistic expression applicable
       to large v see Ch. 9.)

3 K etc.   temperature in degrees on the absolute scale.

fu   flux unit   The unit commonly used in radio astronomy as a measure of
                   the energy per unit bandwidth falling on unit area in unit time.
                   Expressed in terms of watts per square metre per cycle
                   per s(hertz) (W m^{-2} Hz^{-1})

1 fu = 10^{-26} W m^{-2} Hz^{-1}

g   gramme

g cm^{-3} grammes per cubic centimetre

$M_\odot$ mass of the sun ($1 \cdot 99 \times 10^{30}$ kg)

$R_\odot$ radius of the sun ($6 \cdot 96 \times 10^{10}$ cm)

INDEX